CA

Tips and Tricks

Emmett Ross

Author of VB Scripting for CATIA V5

Copyright Information

First Edition – Paperback version

http://www.scripting4v5.com

ISBN-13:978-1500923174
ISBN-10: 1500923176

The author also assumes a general understanding of how to use CATIA V5 including geometry creation and various workbenches (mainly Part Design, Generative Shape Design, and Assembly Design). To learn more about CATIA please refer to the resources in the Appendix for more information about how to use CATIA or where to go to get further answers or advice. I welcome any comments you may have regarding this book. To contact me please email: **emmett@scripting4v5.com**

Your Free Bonus

As a way of saying thanks for your purchase, I'm offering a free resource guide that's exclusive to my book and website readers. Later on in this book I discuss how macros can help you automate your repetitive processes. But writing custom macros is not always easy. That's why I'm sharing with you the seven tools I use when programming CATIA V5 macros. You'll learn how I use each of these tools (most of which are free) and how they help me create macros.

You can download this action-packed PDF file at:

http:www.scripting4v5.com/FREE

Table of Contents

Chapter 1: Introduction

The Author's Story

I wrote this guide because when I was in a time of desperate need, not too long ago, becoming super proficient with CATIA V5 helped advance my professional career. I was working as a new mechanical engineer at a struggling company where everyone else was much more experienced than I was. I realized that made me expendable. When the economy took a turn for the worse, and coworkers began getting laid off, I feared for my job and my family's future. I needed a way to set myself apart to prove my value to the team. Learning CATIA's advanced features and tools gave me a huge advantage over my coworkers. Additionally, once I was able to create macros for CATIA I was able to automate timely processes which helped to quickly earn my colleague's respect, leading not only to me keeping the job but also to quicker promotions, along with more job freedom and flexibility. Not only did it help me bounce back from a low point but it opened my eyes to the world of productivity, efficiency, and automation and the opportunities that it can create for anyone's career. And if I can do it then you can too!

CATIA V5 Tips and Tricks is my way of giving back for all of the fortunate things that have happened to me ever since. This is a guide, and the purpose of this guide is to do just that - guide you. If I can help just one person learn one thing that will help them in their career and/or life, all the time and effort I have put into writing these CATIA tips will have been totally worth it!

Why use CATIA V5?

Why use CATIA V5? With its wide variety of extreme uses, CATIA V5 is the Swiss Army Knife of CAD software packages. Simply put, it's the best one available today. It doesn't matter what your skill level is - anyone can learn to use CATIA! Millions of designers and engineers use CATIA V5 everyday but a vast majority of them have not unlocked the full potential of this incredible program. In fact, it's said 80% of the users only use 20% of the features. The purpose of this text is to increase your knowledge in order to improve your productivity and efficiency. If you've ever thought to yourself "there has to be a better way to do this," you're probably right and this book is for you.

Expectations

This book is not a beginner's guide nor is it for those who have never used CATIA before. A basic understanding of the software is required. The purpose is to share tips and tricks you can put to use and get results immediately. The text was written primarily for CATIA V5 for Windows, R20. Most tips should work for all R versions but please be aware some may not. All the tips have been grouped into different categories but besides that are not listed in any particular order. The table of contents can be used as a quick index and there is a list of resources in the Appendix located at the back of the book.

Chapter 2: Setting up for Success

Tip #1: Common Keyboard Shortcuts

CATIA V5 is a powerful tool, but you can't call yourself a power user until you've mastered the essential keyboard shortcuts – it's one of the best ways to easily increase your productivity. Listed below are the CATIA shortcuts you should get to know.

Esc: Abort the current process or exit the current dialog box (when there is one)
F1: Get CATIA V5 assistance by launching the contextual online help
F3: Toggle specification tree display on and off
SHIFT - F1: Context assistance (Get help on toolbar icons)
SHIFT - F2: Toggle the specification tree overview on and off - opens an overview on specifications tree in a new window.
SHIFT - F3: Structure tree activate around e.g. character size to modify (activate the graph if the model is active and inversely)
SHIFT + left: rotate to the left
SHIFT + right: rotate to the right
SHIFT + UP: rotate upward
SHIFT + down: rotate downward
ALT + F8: Macros start
ALT + F11: Visual basic editor
CTRL + PAGE UP: ZOOM IN the model or tree whichever is active
CTRL + PAGE DOWN: ZOOM OUT the model or tree whichever is active
CTRL + RIGHT arrow: Move the model to the right (PAN right)
CTRL + LEFT arrow: Move the model to the Left (PAN left)
CTRL + TOP arrow: Move the model to the TOP (PAN Up)
CTRL + BOTTOM arrow: Move the model to the Bottom (PAN Down)
CTRL + SHIFT + Right arrow: ROTATE the model around z axis Clockwise

CTRL + SHIFT + LEFT arrow: ROTATE the model around z axis Anti Clockwise
CTRL + Tab: switch between the different windows (swap active document windows)
CTRL + N: New document open
CTRL + O: Document open
CTRL + S: Document save
CTRL + P: Document print
CTRL + F: Search
CTRL + U: Update
CTRL + X: Cut out
CTRL + V: Insert
CTRL + Y: Redo
CTRL + Z: Undo
Up arrow: Relocate the graph 1/10th (one tenth) of a page to the top
Down arrow: Relocate the graph 1/10th (one tenth) of a page to the bottom
Left arrow: Relocate the graph 1/10th (one tenth) of a page to the left
Right arrow: Relocate the graph 1/10th (one tenth) of a page to the right
ALT + Enter: Properties
ALT + SHIFT + Right arrow: ROTATE the model
ALT + SHIFT + Left arrow: ROTATE the model

Tip #2: Understand UUID

Every CATPart and CATProduct contains a UUID, a **Universal Unique Identifier**. Basically, CATIA identifies files based on their file name and their UUID. Where problems occur are when two pieces of data have the same UUID. Compounding the problem, the UUID can't be viewed or edited with any current CATIA function. There are cases when two files may have different names but share the same UUID. This causes a problem when dealing with **Product Data Management (PDM)** systems, like SmartTeam. It is recommended to create new UUIDs whenever possible. Actions which will **create new UUID** include:

File + New
File + New From
File + Save As - option save as new document
INSERT New Product
INSERT New Part
Document Template Creation

Actions that will **keep the same UUID** for each include:

File + Open
File + Save Management
File + Save
File + Save As
Send to directory
File + CLOSE
File + Save
File + Save ALL

Tip #3: Import toolbars from one workbench into another

You can save time by setting up custom icons for Commands and macros. If you place these icons on a toolbar specific to GSD, then when you switch to part design you'll have to setup the icons on another toolbar all over again. Save time by importing toolbars from one workbench into another. Select the **Customize** option from the **Tools** pull-down menu and the **Customize** dialog box is displayed. Select the **Toolbars** tab from the **Customize** dialog box. The options available in the **Toolbars** tab are displayed. Choose the **New** button from the **Toolbars** tab of the **Customize** dialog box. Now, select a workbench from the list of workbenches and a toolbar from the toolbars area. You can use this to import toolbars from any workbench.

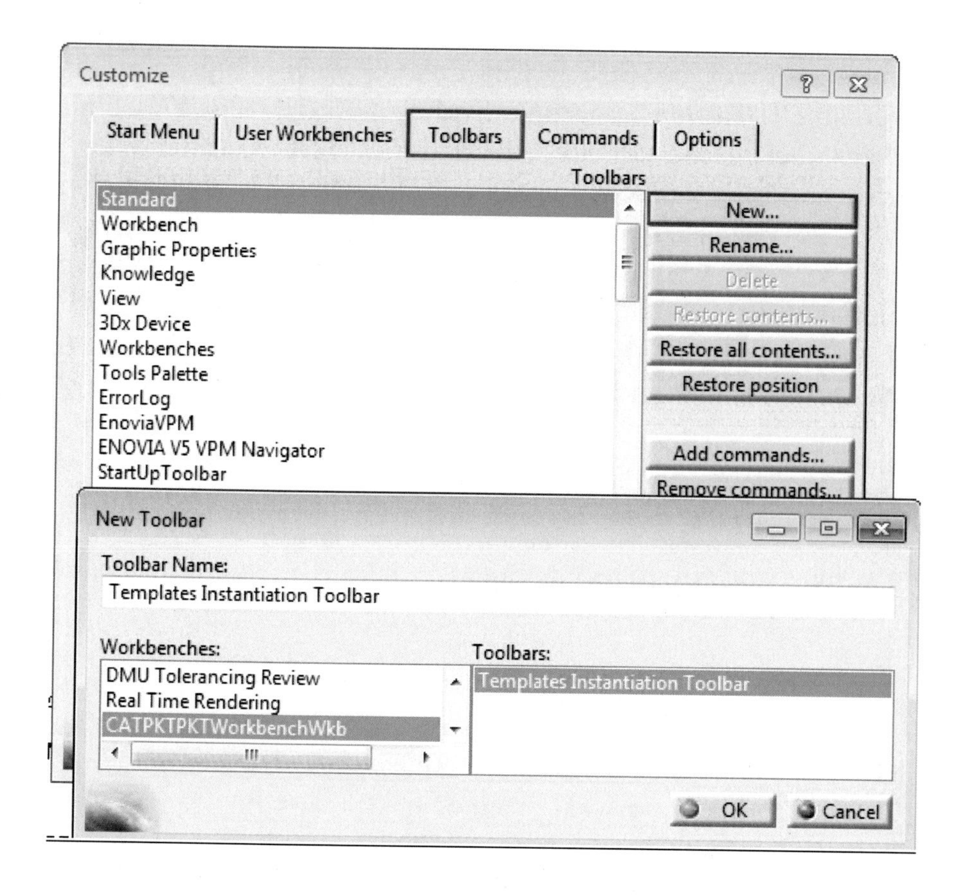

Tip #4: Check CATIA release version

To check the release version of a CATPart from within CATIA go to **File > Document Properties**. There you'll find information about the saved document version. For example, 5.22.4 means CATIA V5 R22 SP4.

Sometimes you may try to open a file in CATIA but it gives an error because it was created with a higher release version. There are programs available to download that will check the release version outside of CATIA. Another way to determine the version, release and service pack of a CATPart is to go to the directory where the CATPart is saved, right click on the CATPart, and open with either WordPad or Notepad. Perform a "Find" on "**minimalversiontoread**" (notice there are no spaces). You'll get the information as shown but please keep in mind this method will take longer for larger part files.

Tip #5: Using Shareable Licenses

If you're using sharable licenses, such as Functional Tolerancing (FTA), GSO, etc. then all users should pick the licenses from the "Sharable tab" Under **Tools>Options>General.** That way it keeps the license from being a "default" every time CATIA is opened. It also allows users to add that sharable license without restarting CATIA like you do when you pick it from the licensing tab and you end up having to restart to gain that ability.

Tip #6: Manage CATSettings

CATSettings are constantly written and can easily get corrupted so it's a good practice to backup your CATSettings directly through Windows. Manually browse to the location where the CATSettings are saved in based on your environment, typically something like one of these:

C:\Users\(userlogin)\AppData\Roaming\DassaultSystemes\CATSettings

C:\DocumentsandSettings\user\ApplicationData\DassaultSystemes\CATSettings

Create a copy of the CATSettings folder and rename it with the date: *CATSettings Backup 8-1-2014.* When unpredictable things happen, replace your settings with a clean set. Re-installing CATIA does not clean up your settings directory.

It's also recommended to keep a separate CATSettings folder for each release of CATIA because it is forwards compatible but not backwards. Do not use the same settings folder for multiple versions (R18/R19); I would create CATSettingsR18 and CATSettingsR19, to do this you need to modify in the Environment Editor the path in variable CATUserSettingsPath. Go to Tools then Environment Editor. Select the appropriate environment then in the list at the bottom change the folder name at the end of the variable for the CATUserSettingPath (right click and edit). In the settings folder you will end up with a set of directories for each installation of CATIA plus all the backup folders you created:

R20_CATSettings
R20_CATSettings Backup 6-1-14
R22_CATSettings
R22_CATSettings Backup 8-1-14

Tip #7: Clear History

If you're a regular CATIA user you may have run across this error: "*operation cannot be completed because part is already in session. Please close the session to complete the operation.*" This error may occur because of a scenario like this:

A user opens a Product that contains Part A, Part B, and Part C. The user opens Part D, modifies it, and closes it. According to the Desk and Windows, the only file now open is a Product with Part A, Part B, and Part C. But when the user tries to insert an existing part into the Product and browse to Part D the error message described above pops up.

So how do you close the session without having to completely close and reopen CATIA? There is a way to clear the memory cache without having to restart. It's called Clear History. To find the clear history icon follow these instructions: **Tools > Customize > Commands** tab and click "All commands." Drag and drop the Command "Clear History" to a toolbar. The icon looks like a grey gear with a green check mark in front of it. If no history is cached, the button will be grayed out.

Tip #8: Increase the number of Undo levels

The default number of undo levels is ten. In most cases it can be very useful to have a few more, but do remember that this will hog system memory so show restraint when altering this setting.

R16: Tools > Options: Select General > Performances
R18: Tools > Options: General > PCS

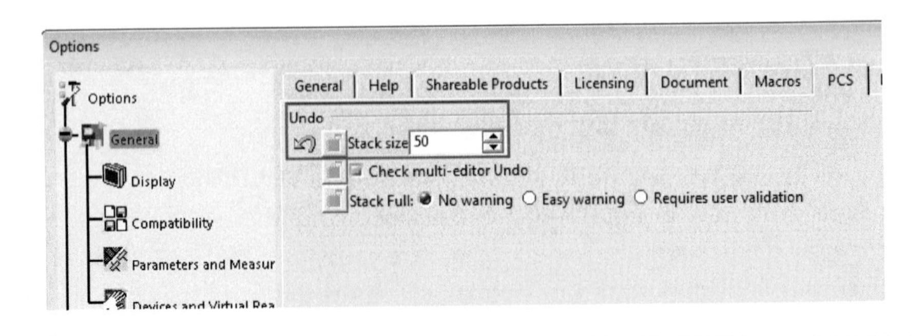

Tip #9: Change the default plane and axis system sizes

Increasing the size of axis systems and planes can make them much easier to see and select. Modify the display size of axis systems and planes by going to **Tools > Options > Infrastructure > Part Infrastructure > Display (tab) > Axis system display size (in mm).**

General
Display
Part Document
Display In Specification Tree
External References
Constraints
Parameters
Relations
Bodies under operations
Expand sketch-based feature nodes at creation
Display In Geometry Area
Only the current operated solid
Only current body
Geometry located after the current feature
Parameters of features and constraints
Axis system display size (in mm) 10
Checking Operation When Renaming
No name check
Under the same tree node
In the main object

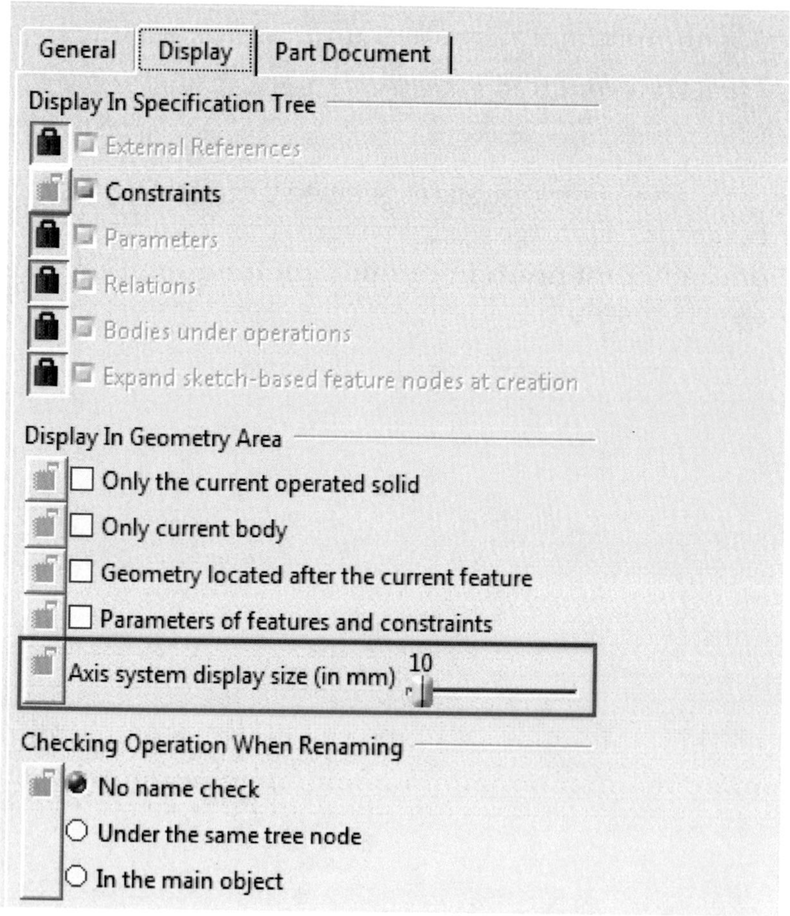

Tip #10: Repeat Actions Automatically

Any command with a toolbar icon can be repeated by double-clicking on the icon. The command repeats until it is clicked again or the ESCAPE key is pressed. The icon will remain orange to indicate that it is repeating, saving the user additional clicks.

Tip #11: Avoid Red

Avoid coloring any geometry or elements the color red because CATIA uses red to 'warn/alert', such as when a part or component needs to be updated. It's a simple way to avoid any confusion.

Chapter 3: Optimizing the Specification Tree

Tip #12: Organize the specification tree

When designing in CATIA V5 you'll doubtlessly be spending massive amounts of time in the specification tree, no matter what you're designing. Therefore, all CATIA users should learn how to build a well named and organized tree and file system. It might seem trivial on small jobs, but on the large ones it becomes an absolute necessity. Label feature names in the spec tree and organize them logically in geometrical sets to make it easier for yourself as well as others to work on your parts.

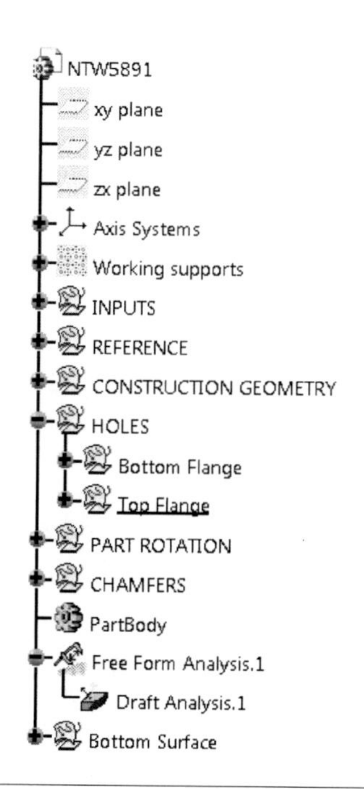

Tip #13: Set the default size for the spec tree

To change the default size of the text in the specification tree in CATIA V5, click on the tree. This will allow you to first set the size of the tree text to how you want it by zooming in or out. Once satisfied, instead of clicking on the tree again, click on the axis system in the lower right hand corner of the screen. This should change the default size of the specification tree, even if you should close CATIA and open it again the text size should remain the same.

Tip #14: Avoid zooming on the spec tree

If you try to click the "+" signs in your specification tree but miss and click a branch instead your model will get a darker shade and all your mouse movements will affect the spec tree instead of the model, an annoying occurrence that slows down your design time. You can avoid this accidental zooming on the spec tree by disabling it. Go to **Tools > Options. Select General > Display > Tree Manipulation** and make sure **Tree zoom after clicking on any branch** is disabled. If you should have the need to resize your tree afterwards, hold Ctrl while scrolling the mouse wheel instead.

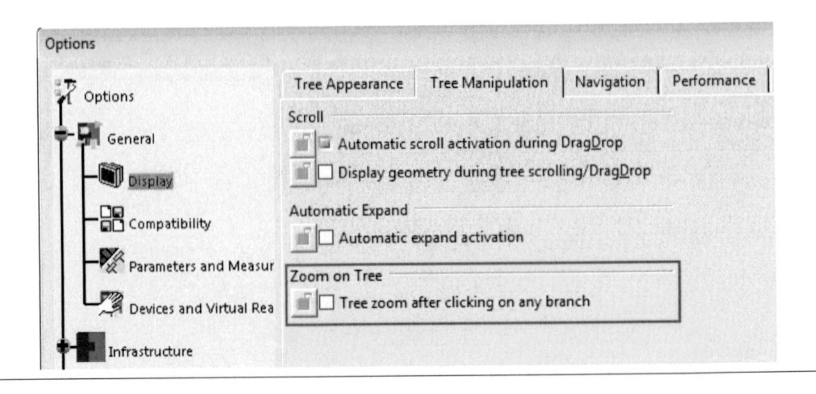

Tip #15: Expand the tree with one click

If you're working with a large part file or assembly you can spend a lot of wasted time scrolling up and down the endless specification tree expanding all the nodes just trying to find a piece of geometry or part file. One trick is to use **Expand** to expand all the nodes in the tree at once. Go to **View > Commands List > Expand**. Or, if you don't want to do every level (which is a bit much most of the time) you can choose to expand only the first or second level. Assign a shortcut or an icon to the command to expand the tree with one click (shown in the next tip).

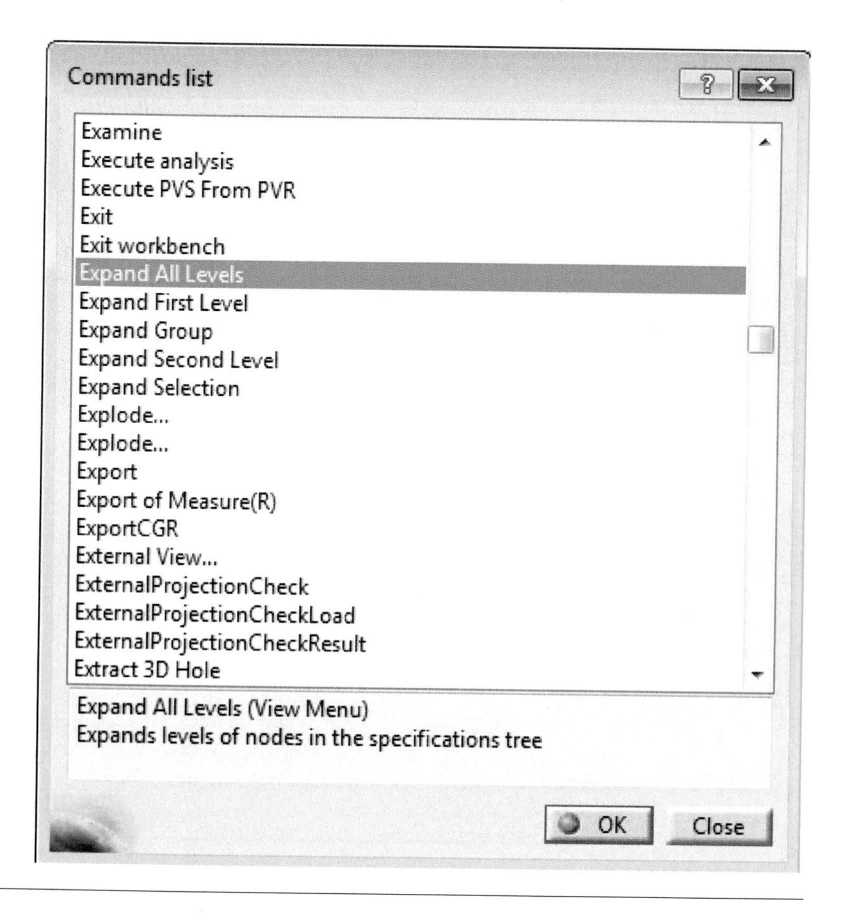

Tip #16: Instantly collapse the tree

Conversely, to collapse all nodes in the specification tree with a single click go to **View > Commands List > Collapse All.** To make expanding and collapsing even quicker, assign an icon to the command by going to: **Tools > Customize > Commands** then select **AllCommands** from the Categories list (at bottom). Select "Collapse All" from the Commands list. Next, select Show Properties. Select an icon button from the list (over 9,000 to choose from)! Select Close. Select Collapse All (text) in Commands list and drag onto the update icon on the bottom toolbar (or any other toolbar you prefer). Once this icon is selected in that workbench, your specification tree will collapse. You need to do this for all the workbenches you use it for. The default shortcut to collapse the tree is Alt+C.

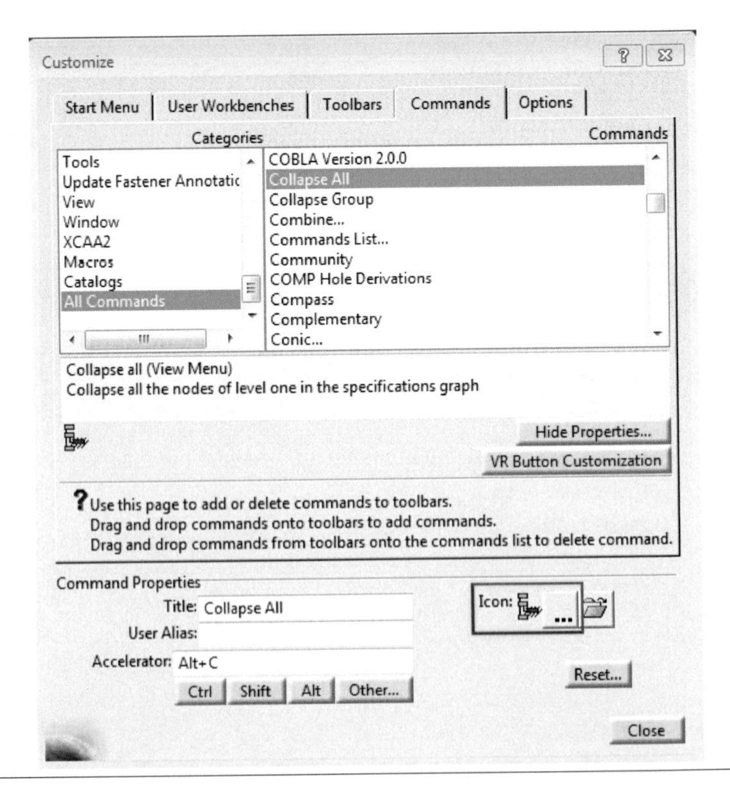

Tip #17: Reorder the specification tree

It's common to not create operations and geometry exactly in the order they will be used. However, there's no need to delete anything, simply reorder the tree. Right Click on a geometrical set and Select **Reorder Children.**

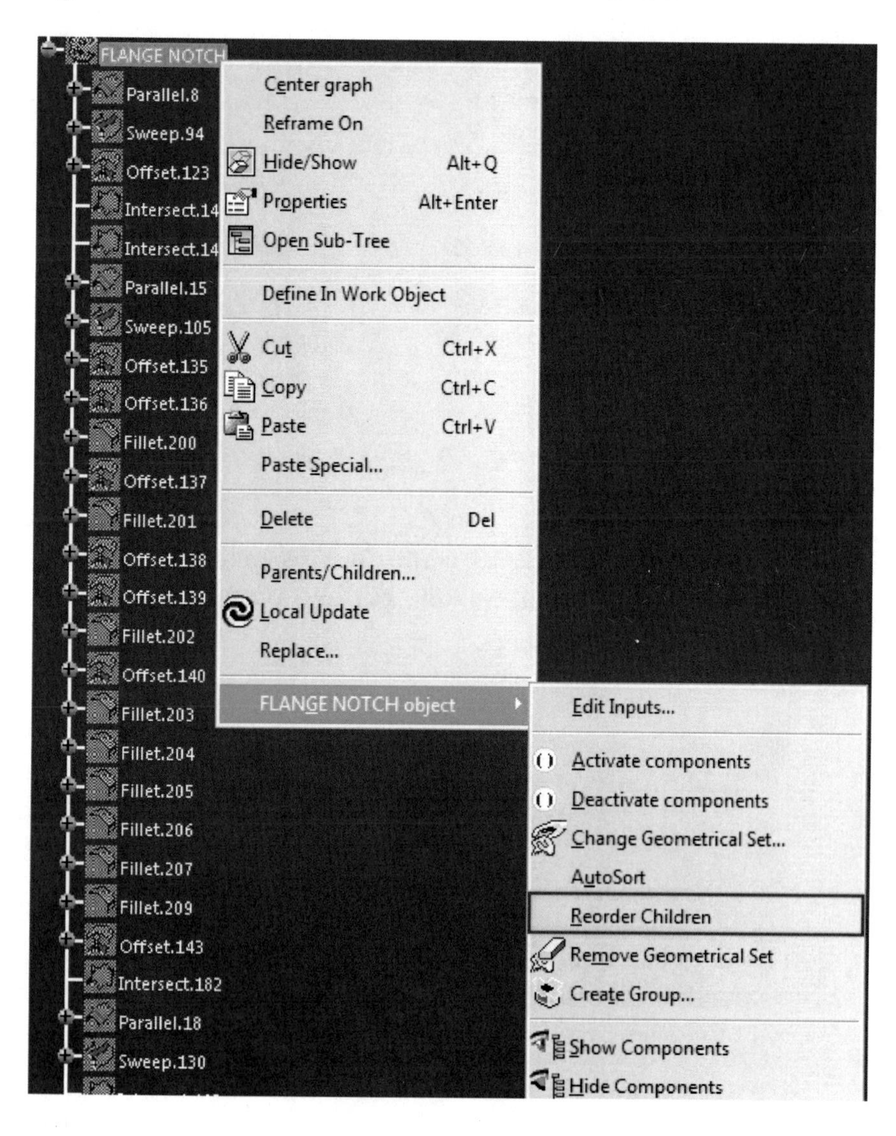

Next, use the up and down arrows to change the location in the tree. Click OK when done.

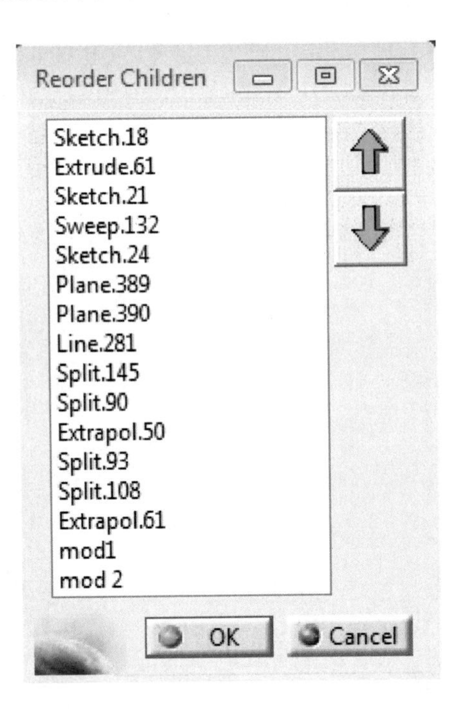

Another option is to create a VBA macro with list boxes where parts can automatically be cut and pasted into the desired order.

Tip #18: Manually export the spec tree

When you want to export your specification tree and you're in assembly design, go to Save As and select TXT file. You will get a text file that contains information only from your specification tree (i.e. part numbers). Another option is to use a CATScript macro, shown in the macro chapter.

Chapter 4: Sketcher

The following tips will help optimize your time and productivity when using the Sketcher.

Tip #19: Change Sketch Supports

Changing a sketch support is useful when you need to change your design, especially since redrawing the model can be tedious and time consuming. The steps to change a sketch support are as follows: Double-click the part for activation, right-click on the sketch, then select Change Sketch Support. The Sketch Positioning window pops up and allows you to change the type and reference geometry for the sketch. Click a new plane to change the sketch support.

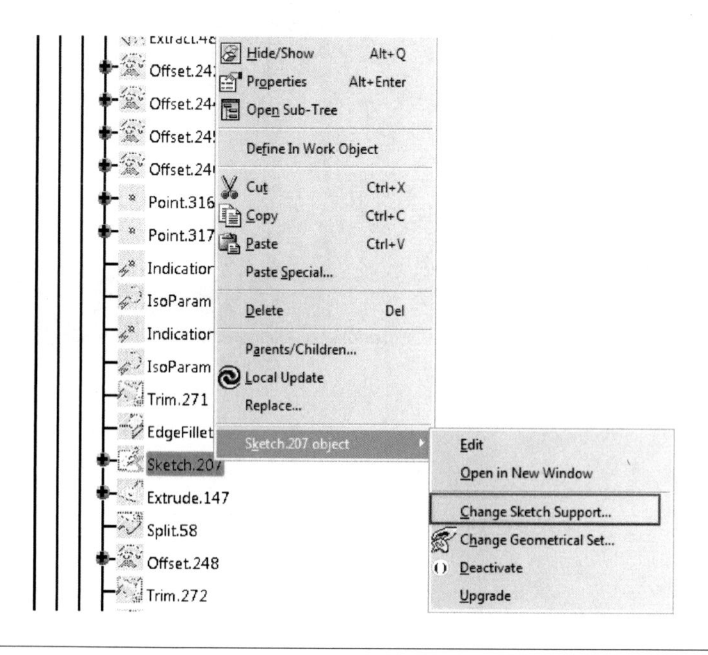

Tip #20: How to scale sketches

Often times, as you're creating a sketch you'll find what you've drawn is either too small or too big, and as you add constraints and modify them, your sketch becomes misshaped. As a result, you need to scale all lines, arcs and circles with the constraints already on the sketch. To do this, use the Scale button but be sure to uncheck Duplication mode. The nice part of doing this is that all the constraints will scale up or down at the same time and in the same ratio. This timesaving trick prevents you from fiddling with the sketch to get it close to the size you need. This is especially helpful on sketches with a lot of arcs and circles with tangencies. It also prevents complex sketches from turning red (with an error), because geometry relations need to increase at the same ratio simultaneously to retain shape integrity.

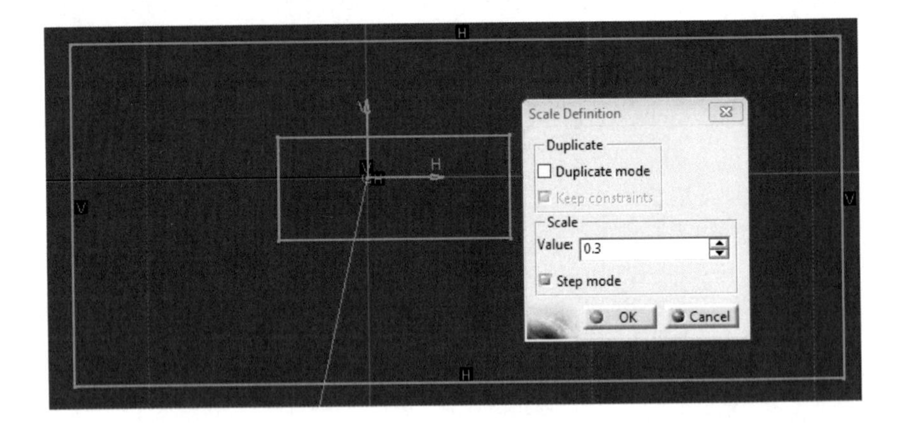

Tip #21: Selection Trap Command

If you attempt to select sketch geometry above an existing solid by clicking and dragging, CATIA, by default, will attempt to Copy and Paste the geometry outside the sketch. To avoid this, click the SELECTION TRAP ABOVE GEOMETRY icon before making the selection.

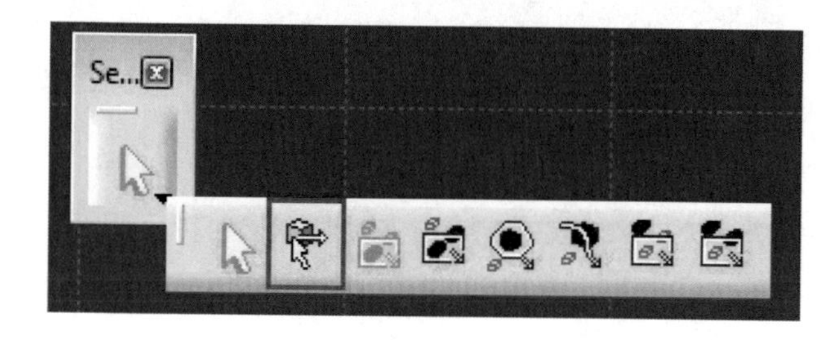

Tip #22: Quickly Constrain a Sketch

The Constraint tool inside sketcher is mostly used to create dimensional constraints between the elements of a sketch. However, it can also be used to create geometric constraints like coincidence, concentricity, tangency, and more. To create a geometric constraint, follow these steps: Click the Constraint icon. Click the first sketch element, and then click the second sketch element. Right-click instead of placing the dimension and then you'll be able to select the intended geometric constraint from a list of options available. This method is much quicker than applying constraints with the Constraints Definition dialog, especially when you combine this with double-clicking to repeat the constraint command.

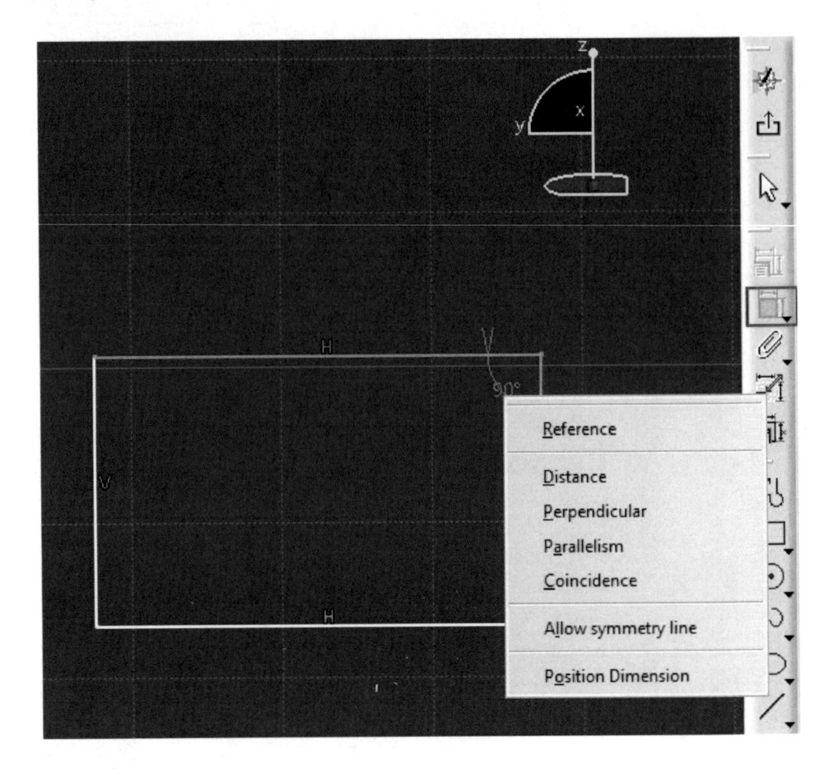

Tip #23: Quickly Select a Sketch Profile

If you have a large sketch profile and you want to quickly select the entire profile, simply select one line or element of the profile, then use **RIGHT-CLICK > OBJECT > AUTO SEARCH** and the entire profile will be highlighted orange.

Tip #24: No Automatic Constraints

While you're sketching in Sketcher, CATIA will automatically try and create geometric constraints based on the geometry you're creating, but often times you don't want the constraints it's trying to make. To temporarily suppress the automatic constraint feature, simply hold down the shift key while you're sketching.

Tip #25: Positioned Sketch vs. Sketch

When creating a sketch there are two main types to choose from: Positioned sketches and regular or sliding Sketch. Positioned sketches allow modification of the location and orientation of the sketch while Sketch allows for the geometry to move from plane to plane if needed. The advantage of using a positioned sketch is the ability to select not only the sketch plane but also the center and the axis directions, and they can all be changed later. Positioned sketches are extremely useful but aren't required in a lot of situations, and if you do use it make sure the sketch is positioned properly.

Tip #26: Make simple sketches

Designers should try to keep their sketches as simple as possible. Except for curves that define major features of a design, it is a good practice to leave fillets out of sketches and add them as dress-up features later. Never duplicate geometry within a sketch. Only draw a single profile; use a pattern to duplicate the profile's features if needed.

Chapter 5: Geometry Creation Best Practices

Tip #27: Use Feature Based Modeling

For light and efficient models use **feature based modeling (FBM)**. Modeling a part in section features allows CATIA to look at the data in clumps and will make open and update times quicker because CATIA can skip ahead when areas are not affected by the change. Feature Based Modeling breaks all joining operations (such as fillets, trim, blends and joins) into two distinct elements: the base and the feature.

Base Element: The fundamental geometry that represents the skeleton of the part, usually the bigger piece of geometry.

Feature Element: Geometric elements that add specific functional characteristics to the part. These are usually smaller geometry and should have minimal inputs.

CATIA follows a specific Selection Order when these elements are combined. The Base element must be selected first and the Feature element second. Feature Based Modeling makes models more flexible and allows you to easily to add and delete geometry.

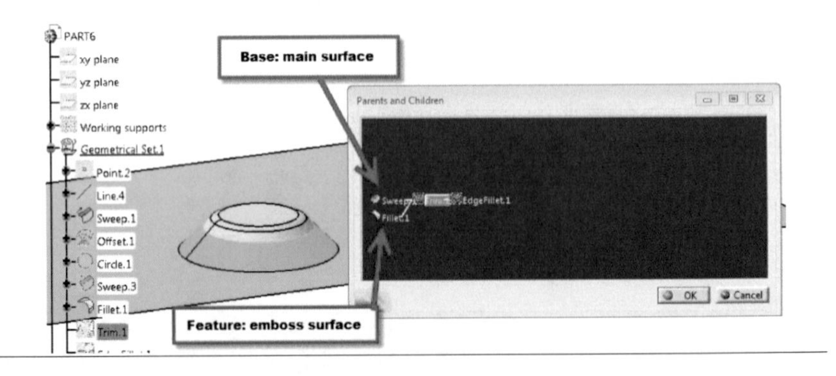

Tip #28: Avoid Selecting BReps

All geometry is made up of a collection of **Boundary Representations (B-Reps)**. Unfortunately, B-reps are not stable elements in the sense that CATIA can't always locate that element again when the geometry is modified. CATIA will search for a similar element in the proximity of where the previous element was located but often times will not find anything or will choose incorrect geometry. Although up front it may save time to create geometry with B-reps, this time will be lost in future updates when the B-reps must constantly be manually selected over and over in order to fix the elements since CATIA cannot find the original B-rep. Sometimes they are unavoidable, because B-reps are required for functions like Edge Fillets and Variable Fillets. Building with B-reps causes more opportunity for the model to fail and updates are inefficient when selections need to be made again and again. Designers should strive to build robust data from the beginning. Therefore, it is advisable to avoid selecting B-reps to create geometry as well as using the extract function to select Faces, Edges or vertices from geometry with a lot of history.

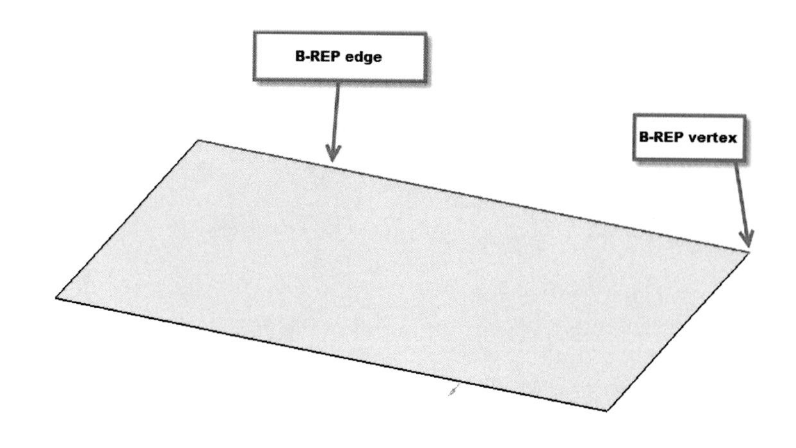

Tip #29: Avoid overusing extrapolate

If there are geometrical elements inside your model, such as Offsets, Parallel curves, etc., that you need to increase in size for an operation later in the tree, then the best practice is to increase the element size at the root element, if possible. Avoid using extrapolate unless absolutely necessary, like if the root element is a dead surface with no history. The fewer elements you use to build your model, the more robust and less complex it will be.

Tip #30: Scaling versus affinity

There are two commands within CATIA's Generative Shape Design workbench which are often confused: scaling and affinity. *Scaling* is used when you want to generate an element scaled equally along the X, Y and Z axis with respect to a reference. On the other hand, *affinity* lets you scale your model with different ratios in the X, Y and Z axis. Utilizing this method gives you more control than using the scaling command. To use these commands, find the operations toolbar in GSD (Generative Shape Design) workbench.

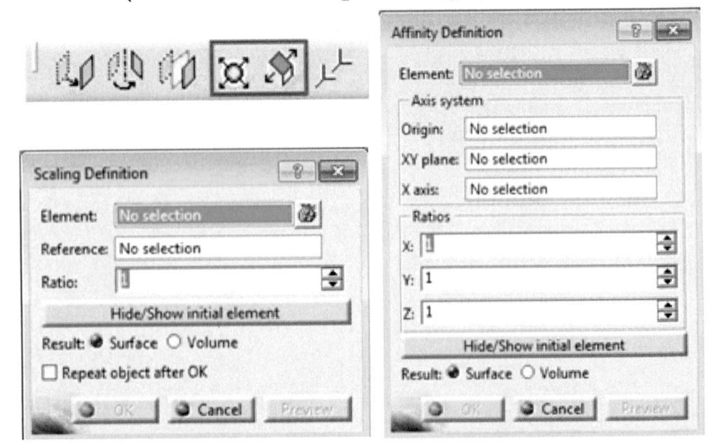

Tip #31: Entering values in any unit

If you type in the correct units for any measurement, it will automatically convert that unit to the default set in Tools>Options. For example, key in 10in and the PAD will measure 254mm. To see what unit abbreviations CATIA uses go to **Tools>Options>Parameters and Measures>Units** (millimeters = mm, inches = in, feet = ft, etc.). You can type in multiple units in one formula and it will convert them all to the default.

Tip #32: Quick Select Tool

The Quick Select is used for obtaining root elements when a piece of geometry gets selected. Quick Select can be found on the User Selection Filter toolbar (Generative Shape Design, GSD, license required). After selecting a feature, the quick select dialog appears showing you the current selection and the parents and children of that object. You can change the current selection by clicking on a different feature on the part, or by selecting on the part or children features in the dialog. Check the Hide other elements checkbox, and then select Parents, Current, or Children to visualize those features even if they are hidden in the specification tree.

Tip #33: Use the magnifier

To see small objects or to zoom in on a specific area, use the Magnifier feature, under the View drop down menu.

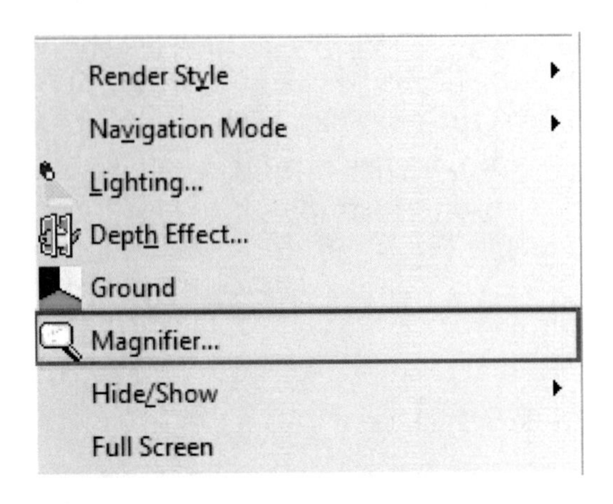

Tip #34: The question mark

A great way to learn about new functions is by using the question mark command, a tool that is directly linked to the help files. To use the command simply click on the arrow with the question mark and then select any item on the screen. This will not only give you a description of the item but will link to the detailed help file page.

Chapter 6: How to Tricks

Tip #35: How to reset the compass

There are multiple ways to reset the compass in CATIA V5. Try any of these methods:

1. Select the View menu from the top then select Reset Compass
2. Place your cursor over the red dot on the compass, the cursor graphic will change to arrows, then press hold the select button and drag and drop the compass with the cursor to the global axis displayed in the lower right corner of the screen
3. In the power input area (lower right corner of the screen) enter "c:reset compass"
4. Finally, create a CATScript macro using this code:

```
Sub CATMain()

CATIA.StartCommand "Reset Compass"

End Sub
```

Tip #36: How to move components in an assembly

Use the Smart Move command to set the initial position of parts in an assembly. Click the Smart Move icon to make it active, then click and drag a part with the mouse to move it around in three dimensional space. The selected part will automatically snap to nearby geometry as it is moving. Check the automatic constraint creation option in the Smart Move dialog box to automatically create an assembly constraint after the part is positioned.

Tip #37: How to measure the length of a spline

Splines are created in Generative Shape Design or in the Sketcher. There are two common methods used to measure the length of the spline if it were straightened out into a straight line. First, you can create a parameter called "Length" and then you would add a formula. Write length() and place the cursor between the parentheses and double-click on the spline. For example: length(`Geometrical Set.1\Sketch.1`). Or, the second, easier method is to simply pick the curve and hit measure item. If you don't get the length of the spline hit customize and set length to the edge group. Now you know several methods of how to quickly and easily measure a 2D or 3D spline in CATIA.

Tip #38: Analyze how a part was designed

Complex parts with numerous features are sometimes difficult to edit without first understanding how they were designed. The **Scan or Define in Work Object** command will allow you to step through a part's structure and visualize its construction. First, go to **Edit > Scan or Define In Work Object...** Next, click First icon to roll back to the first feature that was created. Finally, click the next icon to step through and visualize each feature as they were created.

Tip #39: User selection filters

The user selection filter can help you pick the correct geometric element for the current command and avoid wasted time by picking unwanted geometry and having to re-select. Turn on the user selection filter toolbar in **View > Toolbars > User Selection Filter**. When a filter is active (displayed in orange) only items of that type can be selected. For example, when filleting a padded object, the faces of the pad are not selectable with the Curve Filter active, only the edges are selectable.

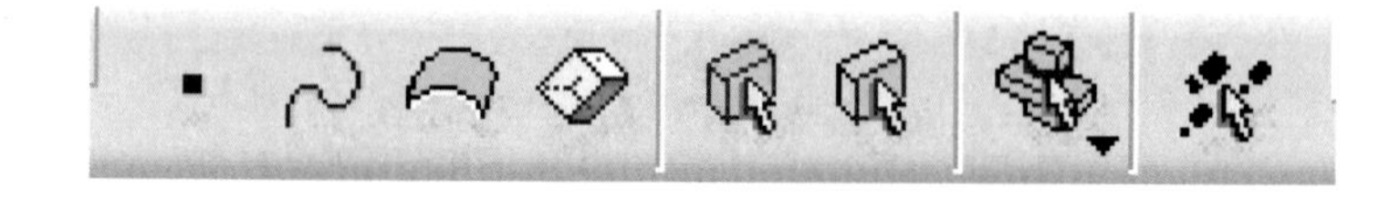

Tip #40: How to create an axis system within a geometrical set

To create an axis system in a geometrical set and not the default axis system node, follow these steps: when you create a new axis, in the dialog box, simply unclick the check box next to the "under the axis system node" option. This will place your new axis under which ever object you have defined as the current in work object.

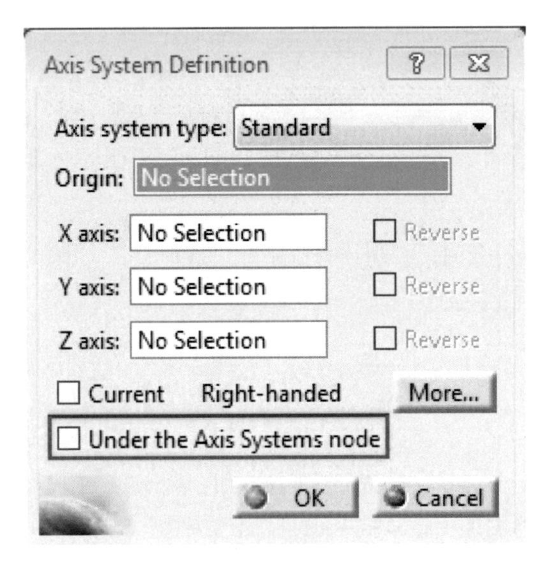

Tip #41: How to duplicate a geometrical set

If you are adding a feature to a part that is going to be repeated, create all the geometry within one geometrical set then use this tool to replicate the entire set and geometry. Go to **Insert > Advanced Replication Tools > Duplicate Geometrical Features Set** then select the geometrical set to be duplicated. Next, select the destination and complete the inputs within the dialog box. Duplicate Geometric Set does a "Powercopy" of another geometric set from within the same part or from one part to another.

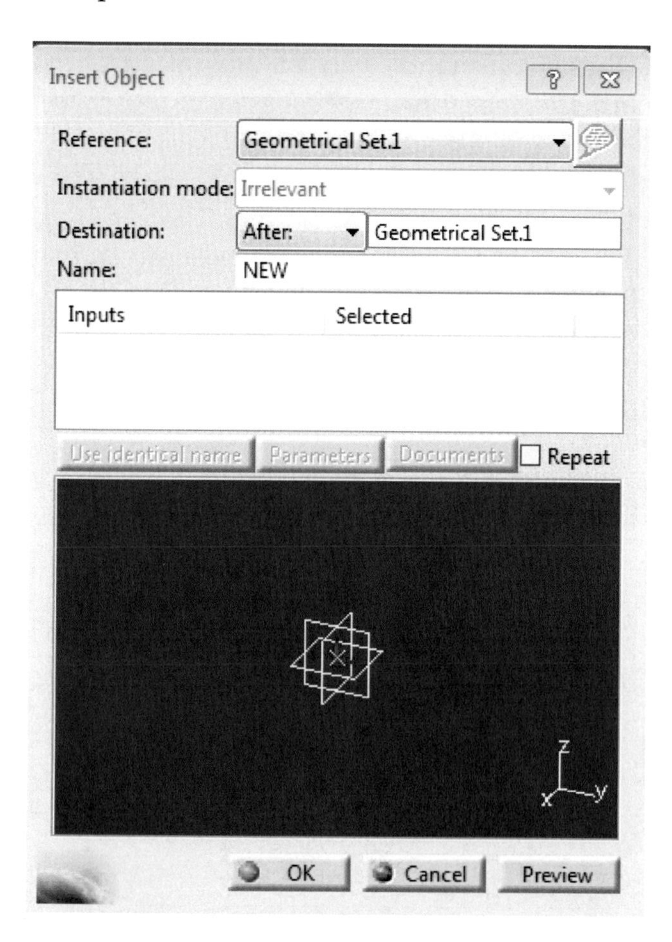

Chapter 7: Tools and Timesavers

Tip #42: Browse the Commands list

To see a complete list of all the commands available in CATIA V5, go to **View > Commands List...** As shown earlier, all of these commands can be linked to a toolbar icon or keyboard shortcut through **Tools > Customize > Commands.**

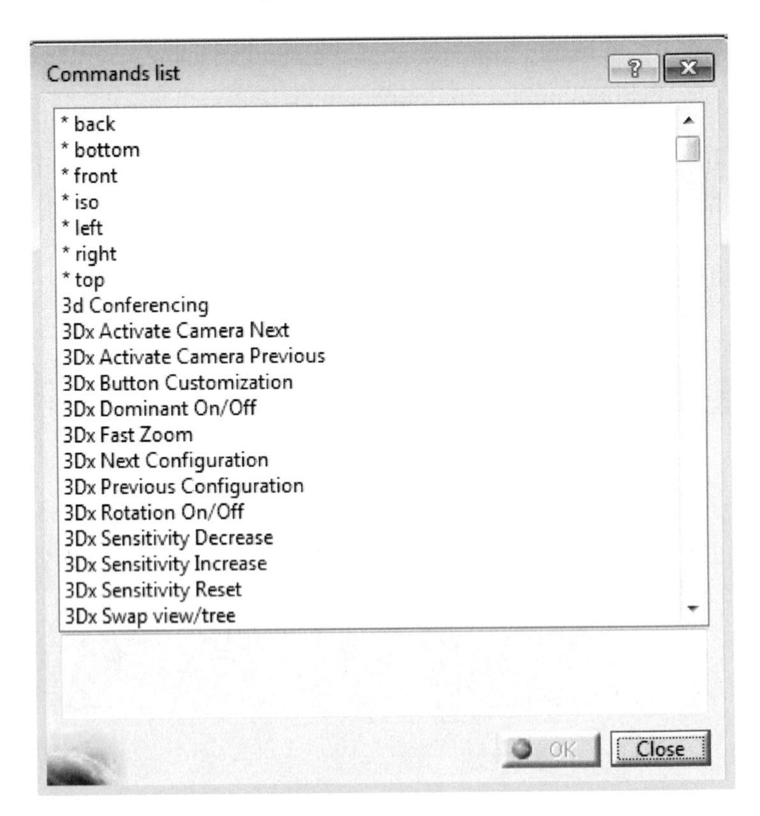

Tip #43: Parameterization analysis

One of the most useful yet underutilized tools in CATIA is parameterization analysis. When a CATPart is the active document go to **Tools > Parameterization analysis**. This nifty little tool allows you to easily list elements in a pop-up window that are important and may be hidden or buried in your spec tree. For example, not only can you can find all sketches, it will also sort by all under, over, or fully constrained sketches!

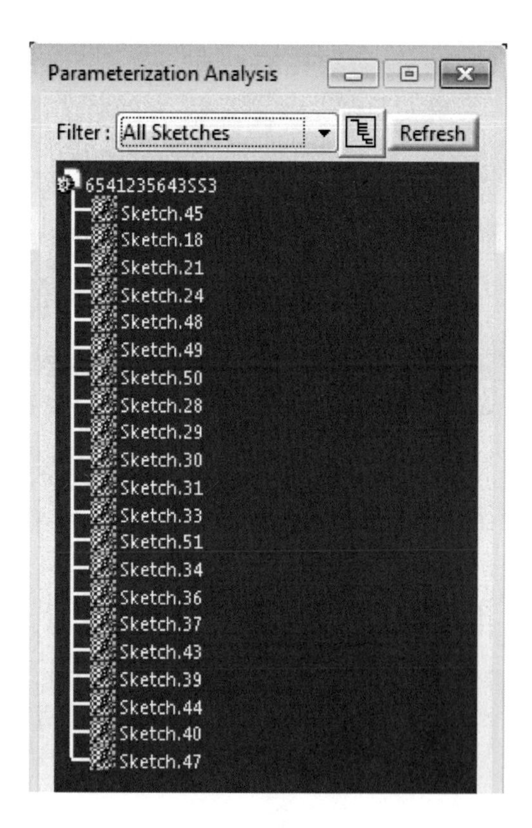

If you've got a large part with lots of surfaces, one way to decrease the file size is getting rid of all the dead end or useless geometry, which can easily be found by doing a Parameterization analysis and filtering by Root Features.

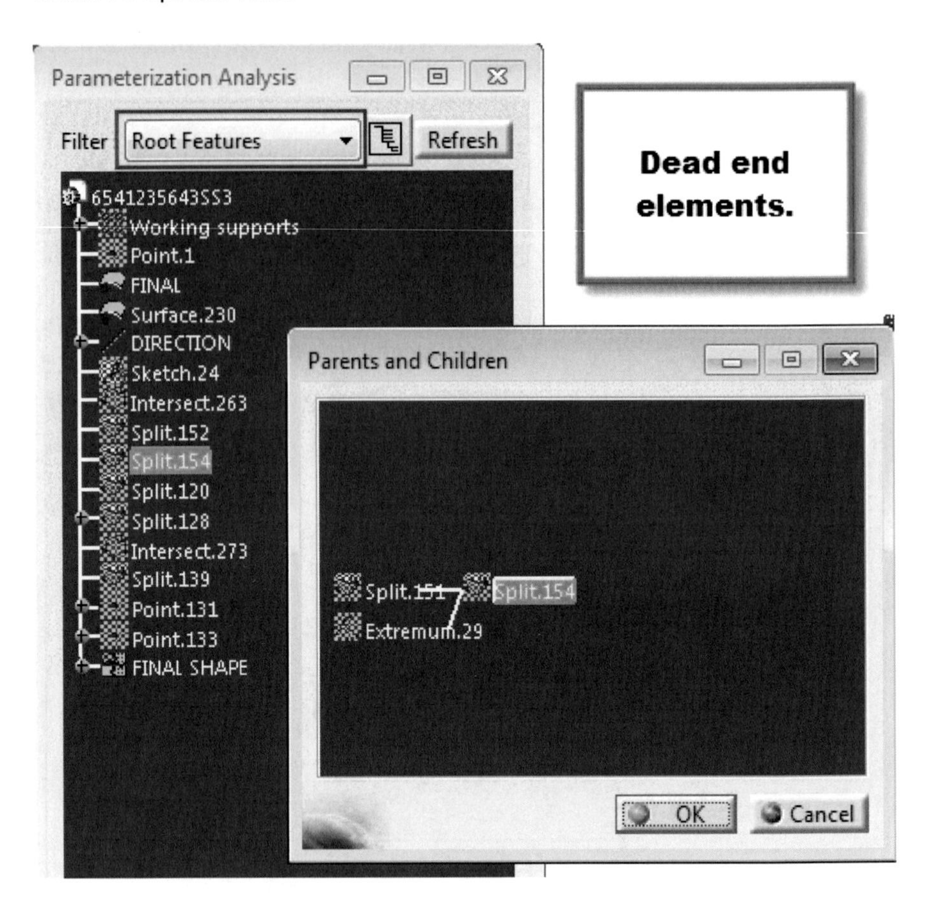

There are multiple ways to accomplish a task in CATIA. The delete deactivate features macro listed later on in this text scrolls through the tree and deletes all deactivated components, but another way to find them is Parameterization analysis and sort by Inactivated Features. You can delete the features from inside the PA window too. Parameterization analysis is a great tool for checking and cleaning up your part files before you submit them to your customer or manager.

Tip #44: Screen capture tool

CATIA has a built in screen capture tool. Access it by going to **Tools > Image > Capture**. Options can be changed inside the tool such as automatically changing the background color to white (useful for inserting into PowerPoint presentations and not wasting as much ink when printing). Of course, this tool is just used for screenshots - taking a snap shot of exactly what is displayed on the screen. For high quality, realistic images you'll need to use rendering.

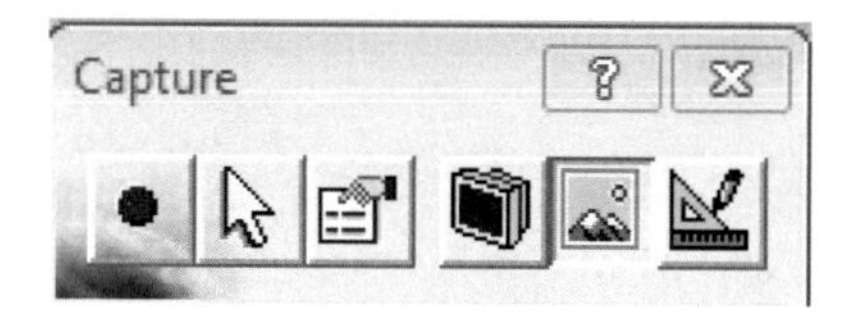

Tip #45: Generate bill of material

To automatically generate a bill of material go to **Analyze > Bill of Material**. Click Defined formats in the bottom right hand corner to add additional properties to display. The BOM can be saved as a TXT, XLS, or html file. BOMs can also be generated automatically from a CATScript macro, either by initiating the Bill of Material tool or by exporting a custom part list directly into Microsoft Excel.

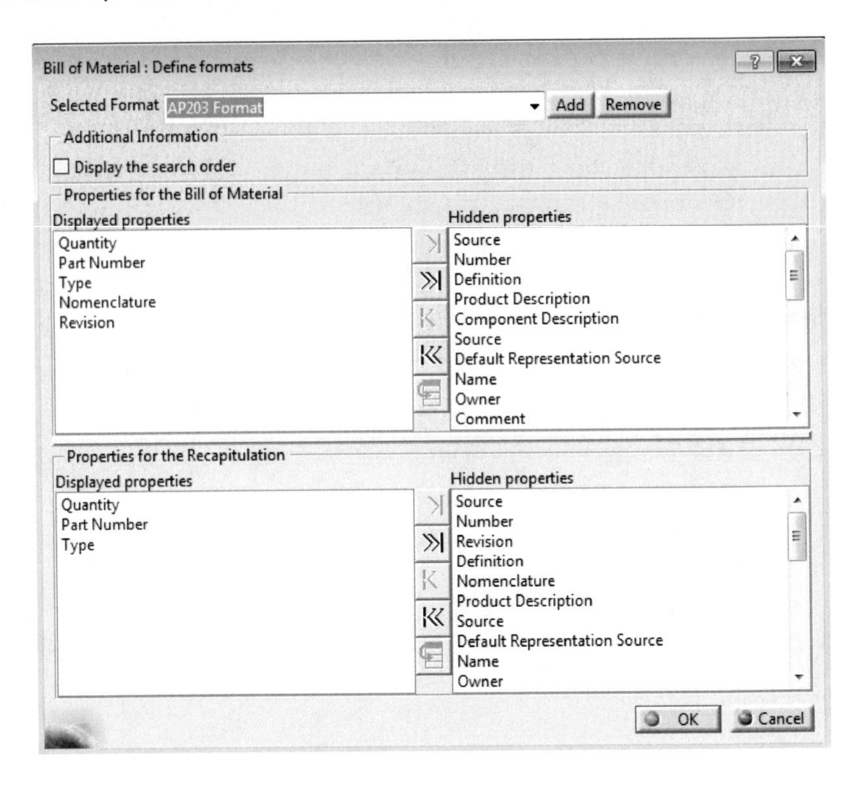

Tip #46: Power Input Box

The Power Input Box is almost always in clear view at the bottom right hand corner of the screen but is seldom used by novice CATIA designers. Instead of clicking an icon, the input box gives the user the ability to directly type in the name of a command or search term to perform a specific action. The default prefix is c: which lets you run commands. To search for objects using your favorite or predefined favorite queries, use f:. The list of useful commands, followed by some examples showing how to use them:

C: name of command
N: feature name
V: search for visible elements, or elements hidden in the No Show space
F: favorite
T: feature type
G: name in graph
NAME IN GRPAH: identical to g: except that the search is case sensitive
S: search for an object belonging to a selection set.
SET: identical to s: except that the search is case sensitive
Col: search for an object of a specific color
L: search for an object located on a specific layer.
D: search for elements using specific linetypes
W: search for elements using specific thicknesses
symb: search for elements with specific point symbols.

A few Power Input command examples:

- "C:Join" will execute the Join command from the generative shape design workbench.
- "n:Point*" will search for and highlight all tree feature that have a name starting with Point.
- "n:healing.3" will select and highlight in your tree and on the model the feature named healing.3.
- "t:fillet" will select all fillets in the part.
- "t:plane & vis:visible,scr" will select only all visible planes currently shown on screen.

Tip #47: CATDUA

CATIA files can accumulate many unwanted elements, known as Ghost Links, due to multiple save operations and modifications during the design process. To get rid of all these undesirable problems the **CATDUA V5 (CATIA Data Upward Assistant V5)** function can be used to quickly clean and optimize your CATIA Files. CATDUA V5 should be used before data importation to ENOVIA or when performance seems to decrease. Running the CATDUA V5 will remove ghost links, reduce file size, and decrease loading and saving time.

CATDUA V5 can be run on a single part through the CATIA Desk by going to **File > Desk**. Always proceed from right to left as cleaning a child part can reveal an error in the parent. Clean only the parts which you can save and beware that Structure Exposed Products cannot be cleaned.

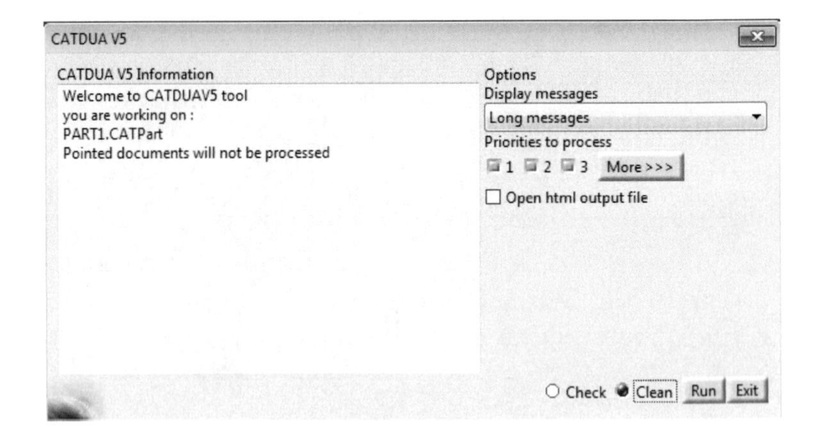

Priority One: performs a cleaning action that may delete data.
Priority Two: performs a cleaning action that may modify data but not delete it.
Priority Three: performs a cleaning action that does not impact data.

Chapter 8: Managing Large Assemblies

Assembly design in CATIA allows engineers to design vehicles with thousands of components. But sometimes even the highest end PCs still crash when loading or editing massive assemblies. The following are tips to make your life easier when managing large assemblies.

Tip #48: Use the Cache System

Using Cache activation will help improve your system's performance dramatically. When this mode is activated, CATIA loads all the parts in assembly in Visualization mode, meaning CATIA doesn't load the entire CATPart with history. Hence it's much lighter on system memory than parts that are put in Design mode.

To activate cache mode go to **Tools > Options > Infrastructure > Product infrastructure > Cache Management tab**. Check the "Work with Cache System" box in Cache Activation. Additionally, you can set the desired path for Cache directory as well as setting the maximum size of the Cache Directory depending on your available free space.

After clicking OK you'll have to restart CATIA in order for the new settings to take effect. The next time you open a CATProduct assembly CATIA will load all the parts in Visualization mode. To switch to design mode in order to edit the components, right-click on the part and select Design Mode.

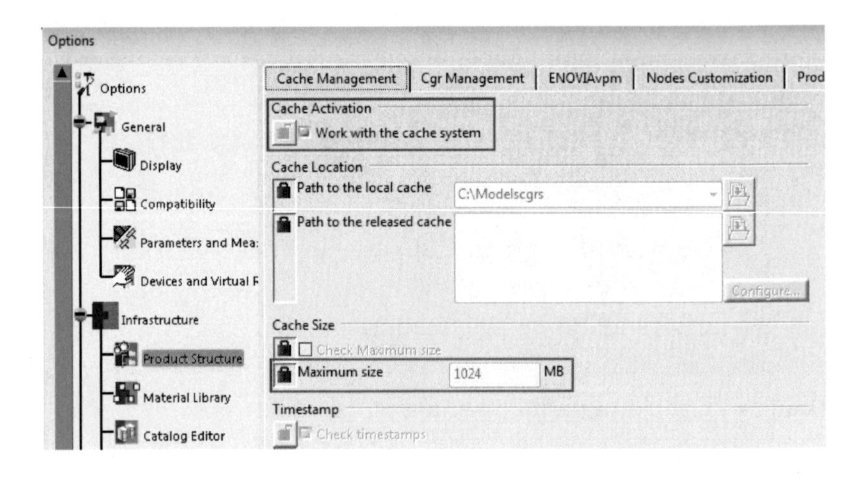

Tip #49: Disable automatic save

By default CATIA automatically backs up the part you're working on by saving it every thirty minutes. The downside is this feature hogs up precious resources and can bog down your machine for a few minutes. To free up these resources, especially when dealing with large assemblies, this feature can be disabled. Disable automatic save by going to **Tools > Options > General** and under Data Save choose "No automatic backup" radio button to disable it.

As stated earlier, reducing the stack size, or number of Undo levels, will also help increase memory capacity and performance. Go to **Tools > Options > General > PCS > Stack Size.**

Tip #50: Optimize CGR settings

Using visualization mode in CATIA versus design mode improves performance by not loading the mathematical data of the part into memory, only a visual of the geometry is loaded into memory. The first time a part is loaded into CATIA using visualization mode CATIA creates a "lighter" version of the model and saves it to the hard drive. Because it creates this light version the first time, it takes longer to load. Successive loads of this model, read the "light" version, which results in a quicker load time and better performance. The file type is saved as a .cgr – CATIA Graphic Representation.

Optimizing the CGR settings will help when managing large assemblies. To tweak the settings go to **Tools > Options > Infrastructure > Product Structure > CGR Management tab**. Check "Optimize CGR for large assembly visualization". CGR files are not downward compatible (CGR files created in V5R14 cannot be read by CATIA V5R13 or lower).

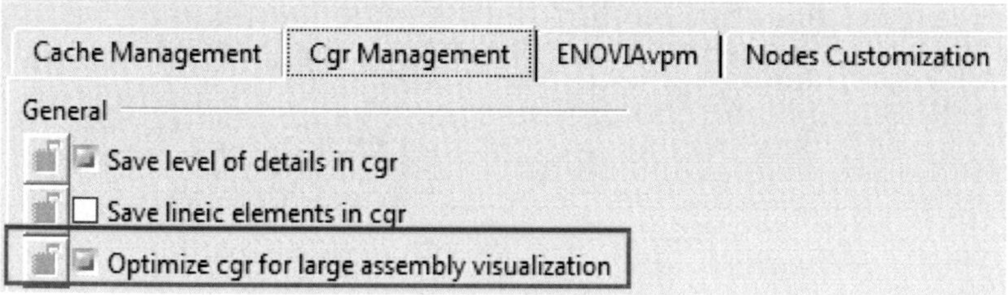

Cache Management | Cgr Management | ENOVIAvpm | Nodes Customization

General

Save level of details in cgr

Save lineic elements in cgr

Optimize cgr for large assembly visualization

Tip #51: Modify display options

Performance can be greatly improved by further tweaking other display options. Begin by going to **Tools > Options > General > Display > Performance tab**. First, disable Occlusion culling by un-checking "Occlusion culling enabled". Occlusion Culling is a technique that does not render objects that are behind other objects to reduce system memory usage. Next, change the 3D Accuracy value e.g. 0.10. Increasing the value increases performance but comes at a cost of the geometry not looking as sharp on your screen. Likewise, increasing the value of Level of Detail while Moving as well as Pixel Culling level While Moving increases performance. Turning off graduated color background also makes a difference in terms of making rendering much simpler as the graphics card will not have to calculate all of the graduated colors that make up the background of the session.

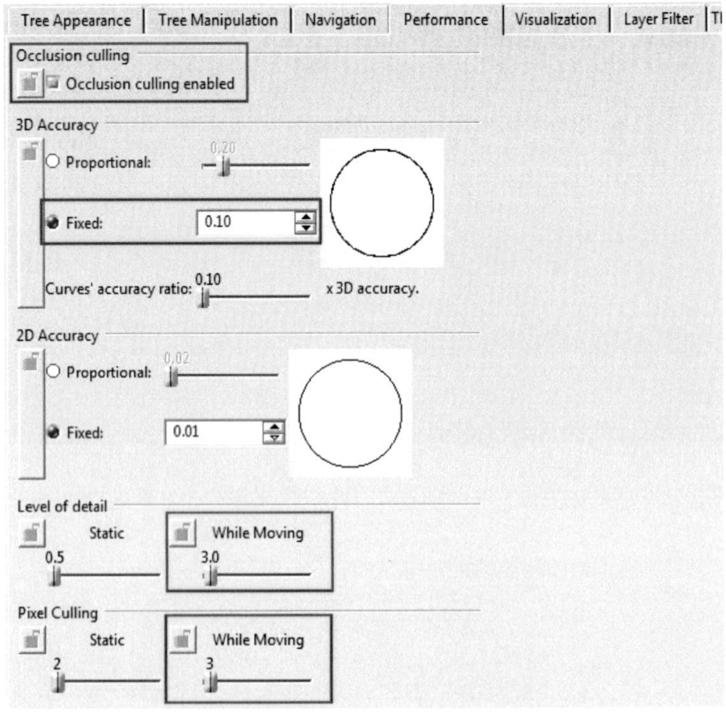

Tip #52: Product visualization

Another method to improve performance when opening a large assembly is to change product visualization setting. Memory usage can be improved by opening assemblies with only the needed components activated and all others deactivated. To change this settings go to **Tools > Options > Infrastructure > "Product Structure > Product Visualization** tab and check the "Do not activate default shapes on open" box and click OK. To improve graphical performance hide components, deactivate nodes, or unload unneeded components.

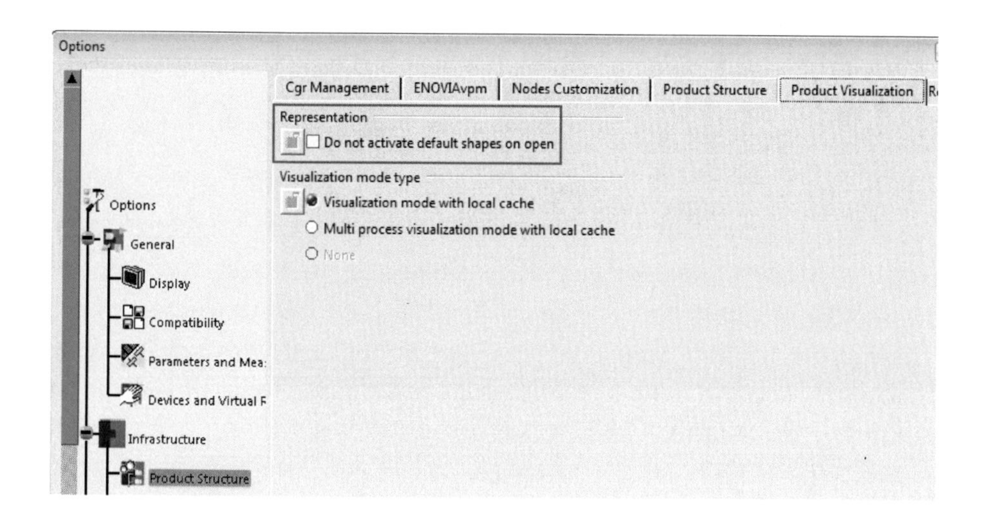

Tip #53: Export as 3DXML

Allow others to see your work for free by exporting as a 3DXML file that can be embedded into a document (Word, PDF, PPT, etc.). The end user just needs to download the free 3D xml player. To export a file as 3D xml click File >Save As > then select ".3dxml" as the file type. 3D xml settings can be changed by going to **Tools > Options > General > Compatibility > 3D XML tab.**

These 3dxml files are extremely lightweight and can be embedded in every type of office document. To embed into a Word or PowerPoint file use Insert > Object. Save as usual. After end users have received the file and downloaded the player, open the document and double click the Dassault logo to view the 3D model. Get the free player here: http://www.3ds.com/products/3dvia/3d-xml/1/

Tip #54: Generate CATPart from Product

If you have a large assembly and need to transfer the data to someone else at a different location to review, one option is to turn all the parts of a product into bodies inside a single CATPart, one CATProduct at a time. To do this, go to **Tools > Generate CATPart from Product....** Give a name as desired for the generated single CATPart. The command will capture all geometrical representation detected in all activated nodes that are children of the selected node. Each geometrical representation found will be created in the CATPart as an isolated body. Elements in No Show are not converted. The Merge all bodies of each part in one body option allows you to merge, for each part, in one body all its part bodies through an add operation and all its Geometrical Sets through the Change Geometrical Set command.

Chapter 9: Drafting

Tip #55: How to reorder drawing sheets

Often times when dealing with a large CATDrawing with multiple sheets, the sheets are typically not in any logical order, which can get quite confusing. The best thing to do is rearrange the drawing sheets in a logical order, and not just rename them. If you want to reorder the sheets in V5 R17 or lower then you are stuck doing it the old fashioned way: cut and paste. Don't worry, this method should still keep any links to CATParts or CATProducts, simply select the sheets in the order you want, cut, and then paste as specified.

Fortunately, if you're using R18 or higher you now have access to a Reorder tool. Select the drawing sheets you want to rearrange, right click, go to Selected Objects, then click Reorder. If you hover over the sheets an arrow will appear.

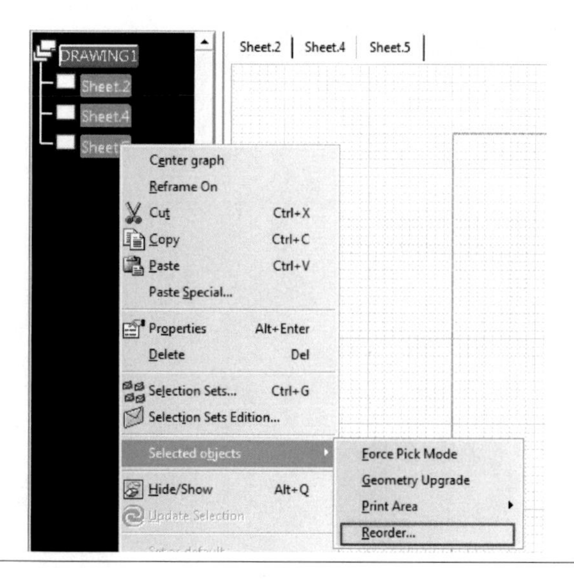

Tip #56: Add trailing zeroes to dimensions

Add trailing zeros after integer dimensions (which cannot be obtained by changing the precision value) by going to **Dimension Properties > Value** tab then change the Description Property to ANS.DIMM.

Tip #57: Resize hole axis

After the creation of an axis system or a thread using the "Axis and thread" functions it is possible to resize this axis system as you'd like to. By default, if you drag one of the extremities, both axis will be resized. However, by holding the Ctrl button and dragging one extremity will only change the size of the selected axis line.

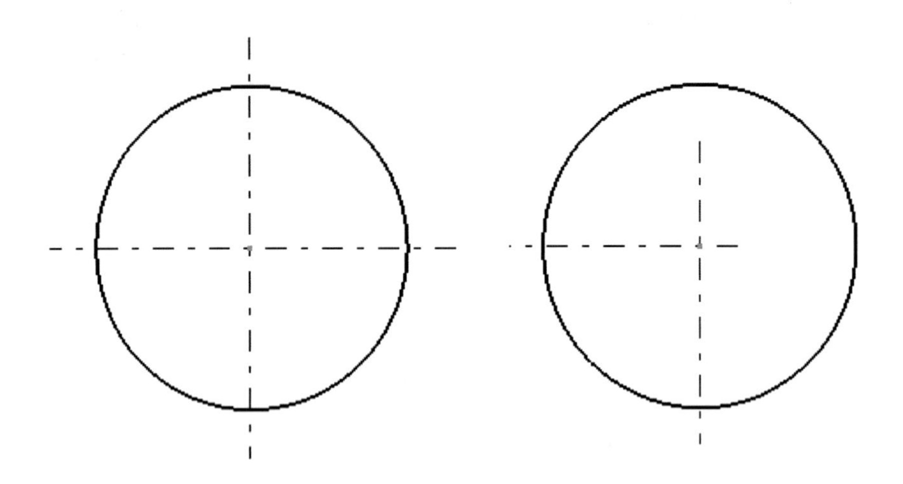

Tip #58: Toggle annotation position snapping

By default, dimensions and annotations on a drawing will automatically snap into position even if the "Snap to Point" option is disabled. To toggle this snapping behavior, hold down the SHIFT key while moving the annotations around.

Tip #59: Export CATDrawing with layers

When a CATDrawing containing layers is saved as a DXF (Drawing Interchange Format) file, by default CATIA puts all the layer items into a single layer and does not export layer names either. There is a setting that can be changed to keep layers intact when exporting from CATIA. Go to **Tools > Options > General > Compatibility > DXF tab** to show all the options for importing and exporting DXF. Under the Export section change the mode to "Semantic" and under Semantic options select "Export layer name." Now CATIA will export layer information from drawings to DXF files.

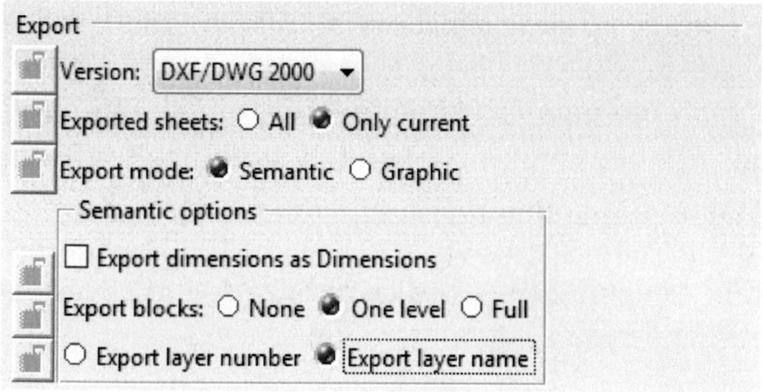

Tip #60: Managing large drawing files - views

The problem of managing a large assembly also results in problems with creating large drawing files. Drawings of migrated models are usually too large and CATIA runs out of memory and crashes, giving the feared and hated "Click OK to terminate" message. There are a number of different options in terms of viewing and settings that will help with large drawing files. There are four different drawing view generation modes: exact view, CGR, approximate, and raster. Each view generation mode has its own strengths and weaknesses.

Exact view is the most accurate and should be used if you're using design mode in the assembly design. Exact view allows you to generate associative dimensions and threads on the drawings from the 3D data. The downside is exact view will use up a lot of memory and is not an ideal option for creating a very large assembly drawing.

Another option is to use **CGR** where only data that is needed is loaded. CGR still allows you to create section, detail, and breakout views as well as generate associative dimensions. CGR mode is solely used for capacity reasons, when memory consumption in exact mode is higher than the authorized allocation on the system. However, if you're working in visualization mode in 3D the approximate mode is a better option over CGR mode.

Approximate mode is similar to CGR mode but only works while the 3D is in visualization mode. But because of this there is no automatic generation of threads, axis-lines, center-lines, dimensions, fillet boundaries, or fillet symbolic representations and you may lose some of the attribute links to the original 3D data. You can further adjust approximate mode by clicking the Configure button in the settings area and change the level of detail to a lower level.

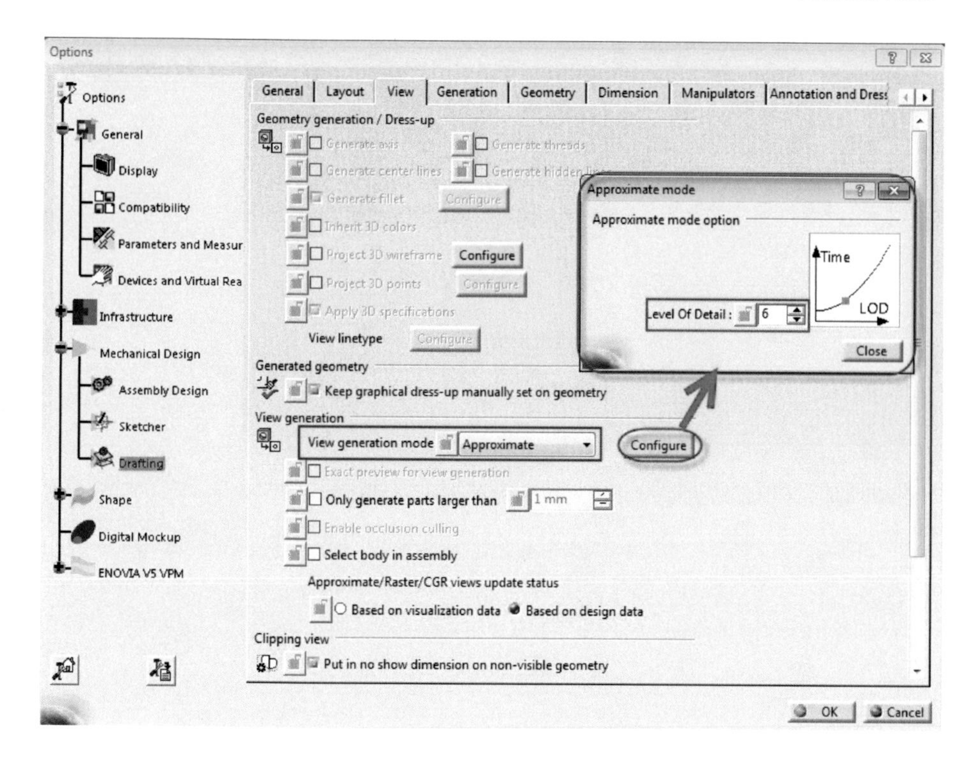

The last mode is raster which can generate very quickly a pixel image of an existing 3D drawing. Shading, color, and even projection and broken views can be created from the raster views. If you click the Configure button after selecting raster you'll see there are various quality settings that allow you to modify the quality of the images according to Dots per Inch (DPI). When printing, it is important to match the DPI rating of the printer in order to get a crisp, clean printing. For example, if the printer prints at 300 DPI, then you would set your view generation to 300 DPI for optimum print quality (**For print > Customize > 300**).

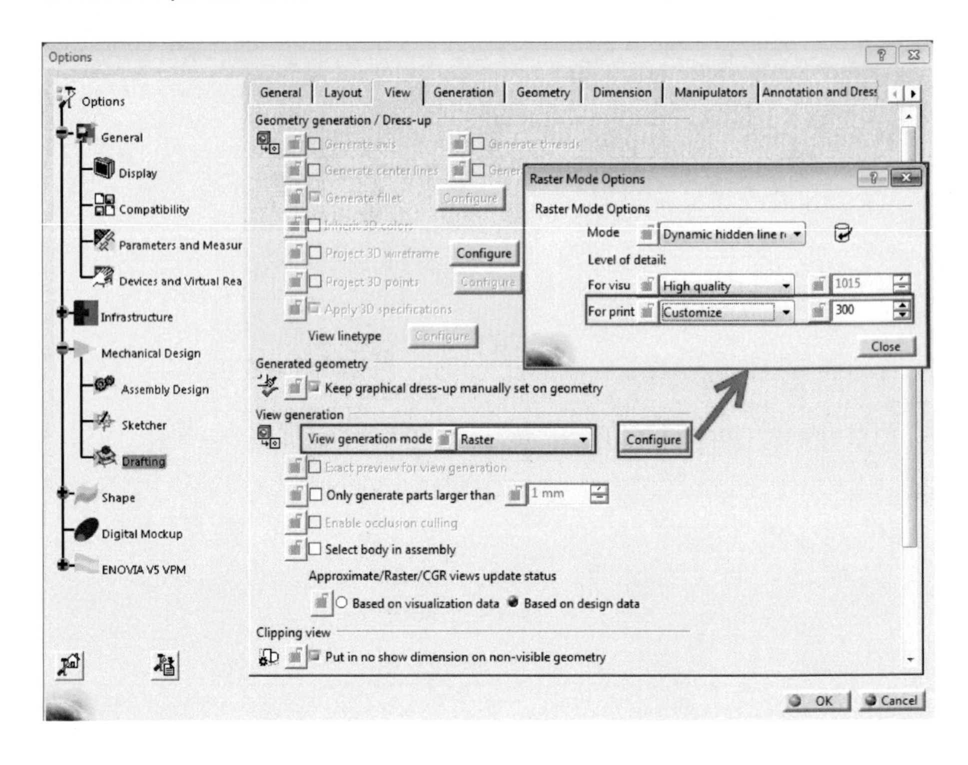

Tip #61: Managing large drawing files - settings

There are other settings that will help manage large drawings. One of those options is to Enable occlusion culling under **Tools > Options > Mechanical Design > Drafting > View tab**. The Occlusion Culling option essentially only calculates the parts that can be seen by the user. This allows the system to avoid un-necessary operations during view update. The option is only available for exact projection mode and it avoids loading geometry of hidden parts (they stay in visualization mode). This leads to significant memory and CPU gains. This was mentioned earlier when discussing large assemblies but this option can also be set as default in tools options, but it is also available as a property for individual views.

Another option is to "Only generate parts larger than X mm". This option can be used with any of the view generation modes and is nice to set to simplify views and reduce processing.

Every time that you want to update a drawing you might want do it view by view and not the entire drawing at once. This is similar to turning manual update on when designing a complicated part and only updating individual features of it at a time. Remember, the less views shown in the 2D, the quicker the drawing will update. If there are parts or features not really needed in the drawing then hide them in the 3D.

Chapter 10: Automation with Macros

A macro is a series of functions written in a programming language that is grouped in a single command to perform the requested task automatically. If you perform a task repeatedly you can take advantage of a macro to automate the task. Why do manual labor when you can simply press a button instead? Macros are used to save time and reduce the possibility of human error by automating repetitive processes. Other advantages include standardization, improving efficiency, expanding CATIA's capabilities, and streamlining procedures. Macros use programming but you don't need to be a programmer or have programming knowledge to use them (though it certainly does help).

The application of automation in the design process is virtually unlimited. Some real world examples of CATIA automation at work:

- Batch script for the conversion of CATDrawing files to PDF
- Batch script to convert CATParts to STP files
- Import of points from an Excel spreadsheet to a 3D CAD model
- Export of data from CATIA model to a bill of material spreadsheet
- Automatic geometry creation from selection
- Automatic drawing creation
- Custom functions

And so on and so on. The possibilities are nearly limitless.

Tip #62: How to create a macro in CATIA V5

There are many CATIA users who want to write macros but simply don't have time to sit down and learn everything they need to know. I will cover those core items to help teach beginners important concepts needed to create custom macros and will show experienced users additional tips and tricks. Macros are created by two primary methods:

1. Macro recorder
2. Write custom code with the macro editor

Once a macro is created, there are multiple ways to open the macros window to run your macros:

1. Go to Tools>Macro>Macros
2. Keyboard shortcut: Alt+F8
3. Assign or create your own icon for each macro

If the macro editor cannot be opened you should talk to your system administrator because it may not have been installed. No extra license is required to run macros, though sometimes licenses for special workbenches are needed if the code uses a function or method from a particular workbench.

CATIA macros are stored in macro libraries in one of three locations: Folders (vbscript and CATScript), Project files (catvba), or CATParts/CATProducts. Only one of these macro libraries can be used at a time. When creating a new macro library, the folder or path location must already exist. If not you will get an error message. Use the following steps to create a new macro library or setup an existing one:

1. Go to Tools>Macro>Macros

2. Click "Macro libraries..."

3. Make sure the Library type is set to "Directories" then click "Add existing library..."

4. Browse to "C:\MyCatScripts" or wherever your CATScripts are saved then click ok or create a new library.

5. Close the macros libraries window. If setting up an existing library you should see a list of .CATScript files. You only need to do this once as the library should load even after restarting CATIA.

Tip #63: Recording macros

One method for creating macros is by recording your mouse actions. For macros recorded in a folder or in a CATPart or CATProduct, Dim statements (declarations) will be recorded for CATScript but not for MSVBScript. For macros recorded in a .catvba library, "MS VBA" is the only choice. Macros **cannot** be recorded while in the drafting workbench. A few things to keep in mind when recording a macro:

DON'T: Switch workbenches while recording a macro.
DON'T: Record more than is absolutely necessary.
DON'T: Use the UNDO button when recording a macro.
DO: Be aware of CATSettings when recording.
DO: Exit sketches before stopping recording.
DO: Check each macro after it's recorded.

Always UNDO what you just recorded and run the macro (you are able to undo CATIA macros after you've run them which is a good way to check if they work as expected or not). If the macro works from within CATIA and repeats what you just did, then the macro obviously works fine. If it does NOT work from within CATIA, you need to fix it. If it does NOT work from within CATIA it will NOT work once you cut and paste it into a VB application.

Look through the recorded macro. Many times extra lines of code are added which are not necessary. This is based on the order of steps that you do as you record the macro. These unnecessary lines can be removed. Recorded macros do not contain any comments or explanations of what is happening in the code and input parameters are never recorded.

For example, a macro is recorded to zoom in and then zoom out it might display the following code:

```
Dim viewpoint3D AsViewpoint3D
Set viewpoint3D = viewer3D.Viewpoint3D
Viewer3D.ZoomIn
Set viewpoint3D = viewer3D.Viewpoint3D
Viewer3D.ZoomOut
Set viewpoint3D = viewer3D.Viewpoint3D
```

Notice how the "Set Viewpoint" command appears multiple times? This is unnecessary in this situation. The viewpoint only needs to be set once after the Dim statement.

Often times you might record a macro with a CATPart active and open it in its own window. All goes smoothly and the macro replays fine. Then, the next day you replay the macro again but this time you may have some other document type open or maybe a part is open but it is in a product assembly. Usually, the macro will fail because when the code was recorded a CATPart was the active document but now it is not. You need to add your own error handling to the code to ensure this doesn't happen. This is one advantage to writing custom code and knowing the fundamentals of CATIA macro programming.

To learn more about programming macros in CATIA V5 and to download some sample code, visit www.scripting4v5.com.

Tip #64: VBA Editor Shortcuts

The following is a list of common shortcuts that can be used in the built-in Visual Basic Editor.

- F1: Visual Basic help
- F2: Open the Object Browser
- F4: Properties Window
- F5: Run macro
- F7: Code window
- F8: Step Into
- Crtl + Break: Break
- Ctrl + J: List properties and methods
- Alt+F11: Go back to CATIA
- End: Quit a running macro

Tip #65: Macro to turn the background color white

To create your first macro after setting up your macro library, go to **Tools > Macro > Macros** and click Create. Change the Macro language to CATScript and change the macro name to background_color.CATScript. Next, copy and paste the code below into the macro editor. Once you've done that, save the macro by clicking save. Close out of the editor. Select your macro from the list and choose Run to watch it turn the background color to white.

```
'This macro turns the background color to white

Sub CatMAIN()

Dim ObjViewer3D As Viewer3D
Set objViewer3D = CATIA.ActiveWindow.ActiveViewer


'change background color to white
Dim DBLBackArray(2)
objViewer3D.GetBackgroundColor(dblBackArray)
Dim dblWhiteArray(2)
dblWhiteArray(0) = 1
dblWhiteArray(1) = 1
dblWhiteArray(2) = 1
objViewer3D.PutBackgroundColor(dblWhiteArray)

End Sub
```

Congratulations! You've created your first CATIA macro! To turn the background color back to the default purple color, use the code below.

```
'This macro turns background color to default
purple

Sub CatMAIN()

Dim ObjViewer3D As Viewer3D
Set objViewer3D = CATIA.ActiveWindow.ActiveViewer

Dim DBLBackArray(2)
objViewer3D.GetBackgroundColor(dblBackArray)
Dim dblPurpleArray(2)
dblPurpleArray(0) = .2
dblPurpleArray(1) = .2
dblPurpleArray(2) = .4
objViewer3D.PutBackgroundColor(dblPurpleArray)

End Sub
```

Tip #66: Macro to Find Geometry

```
'CATScript macro to search for a specific piece of
geometry
'is NOT case sensitive

Language="VBSCRIPT"

Sub CATMain()

Dim oSelection as Selection
Set oSelection = CATIA.ActiveDocument.Selection

Dim iCount
Dim GeoName As String

GeoName = Inputbox("Please enter EXACT name of
geometry to search for.")

oSelection.Search "Name=" & GeoName & ",all"
iCount = oSelection.Count
```

```
'messagebox to display the number of pieces of
geometry found
msgbox "Number of shapes found: "&icount

'loop through all selections
For i=1 to iCount

CATIA.StartCommand "Center Graph"

Next
End Sub
```

Tip #67: Macro to Export the Specification Tree

```
'this CATscript exports the specification tree to
Excel or txt file based upon user input


Language="VBSCRIPT"

Sub CATMain()

Dim productDocument1 As Document
Set productDocument1 = CATIA.ActiveDocument

'Input box to select txt or xls
Dim exportFormat As String
exportFormat = Inputbox ("Please choose format to
export the tree as. Type either 'xls' or 'txt'")

'Input box to enter name of file
Dim partName As String
partName = Inputbox ("Please enter the file
name.")

'Input box to enter file location
Dim oLocation As String
```

```
oLocation = Inputbox ("Please enter the location
to save the file. Example: C:\temp\")
productDocument1.ExportData oLocation & partName &
"." & exportFormat, "txt"

End Sub
```

Tip #68: Macro to delete deactivated features

```
'this CATScript macro deletes all de-active
features (except for sketches)

Sub CATMain()

'error handling
On Error Resume Next

Dim partDocument1 'As Document
Set partDocument1 = CATIA.ActiveDocument

Dim part1 As Part
Set part1 = partDocument1.Part

Dim oDelItem As Selection
Set oDelItem = partDocument1.Selection
oDelItem.Clear

If Err.Number=0 Then

      Dim selection1 'As Selection
      Set selection1 = partDocument1.Selection
      selection1.Search "CATPrtSearch.PartDesign
Feature.Activity=FALSE"

      '---if no deactivated components then end
program
      If selection1.Count = 0 Then
                  Msgbox "No deactivated features."
```

```
      Exit Sub
      Else

Msgbox "The number of deactivated components is: "
& selection1.Count
'---save deactivated features in array
iCount = selection1.Count

    Dim aPoints()
    ReDim aPoints(iCount)

For j =1 to iCount
Set aPoints(j) = selection1.Item(j).Value
Next
'--------------------
selection1.clear

For i=1 to iCount

oDelItem.Add aPoints(i)
CATIA.StartCommand "Center Graph"

If MsgBox ("Deactivated component is: " &
aPoints(i).name & ".  Click yes to delete or click
no to continue.", vbYesNo) = vbYes Then

oDelItem.Delete
End If

oDelItem.Clear
part1.Update
Next
End If

'---error handling---
Else
     Msgbox "Not a part document! Open a single
part document."
End If
End Sub
```

Chapter 11: Troubleshooting

You're trying to create a CATIA part or write a CATScript macro but now you're simply stuck and don't know what to do next. Or maybe it's just not working the way you intended it to. How in the world are you supposed to go on from this point? If you're new to CATIA V5 you're likely going to be dead in the water until you get some help. Before giving up, there are a few steps I recommend you take to try and figure the solution out for yourself. I strongly believe you learn more through struggling, overcoming obstacles, and doing it yourself. Here is my list of actions you should take if you consider yourself stuck:

Tip #69: Use Previous Examples

A great way to learn is by dissecting templates or macros to see how other users solved similar problems. Remember to check for things like parameter and relations. Forums such as Eng Tips and COE are also a great place to see examples. Posting in the recommend forums listed in the resources at the end of this book is a great way to get feedback from multiple power users. Use the CATIA Help files for more examples.

Tip #70: Step Away From the Problem

This is actually my favorite tip in this section. There has been countless times where I've been banging my head against a wall, not able to figure out a problem. So, I would simply get up from my computer and walk away, maybe for a few hours or days, and not think about the problem at all. Then when I sit down in front of the screen again refreshed the answer hits me almost immediately. Seriously, this happens almost every time! It's that whole not being able to see the forest through the trees type of thing. So take a break!

Tip #71: Use the Internet to Ask Questions

Google is your friend! If you don't find a suitable answer on your first search try a new one with different words or a different search engine. Personally, there's nothing that annoys me more than when someone asks me a question and I simply Google it and I find the solution right away. Don't be lazy! Other people do not want to do your work for you.

Chapter 12: Final Advice

Tip #72: Career Advice

What are the best CATIA skills to learn? In my professional career I have found there to be a few CATIA skills that, if you possess a few or all of them, will make you a nearly invaluable and indispensable resource. My top four CATIA skills to learn are:

- Advanced Surfacing (Generative Shape Design workbench)
- NC Programming (Advanced Machining workbench)
- Stress analysis/FEA (Generative Structural Analysis workbench)
- CATIA VBA Macros

Don't be afraid to ask your employer for training in a specific area. Remember to show/tell them how it will improve your efficiency and productivity.

Good Luck!

Our time together is nearly complete. At this point, you should have an idea of what it takes to become a CATIA V5 power user. Plus, you now have over 70 tips and tricks to impress your coworkers or class mates. Remember, *"Little hinges swing big doors."* Put all these tips to work together and they will open up powerful doors in your career.

CATIA V5 is a Swiss army knife; from advanced machining to ergonomics, designs cars, boats, or airplanes, CATIA can do it all. It takes an experienced user to do so, and the more familiar you are with all of CATIA's features, the more quickly your tasks can be accomplished. There are multiple ways to do things but one of those ways is probably better than the others, you just have to figure out which method that is. Saving a little bit of time here and there can add up over a long period of time, such as your professional career. I hope this guide has helped you find a better way to do a few of your daily design tasks.

Thanks for reading and here's to your success!

-Emmett Ross
Author of VB Scripting for CATIA V5: How to Program Macros
http://www.scripting4v5.com

Would You Like to Know More?

Would you like to know more about how CATIA works? Are you tired of repeating those same time-consuming CATIA processes over and over? Worn out by thousands of mouse clicks? Don't you wish there were a better way to do things? What if you could rid yourself those hundreds of headaches by teaching yourself how to program macros while impressing your bosses and coworkers in the process?

My book, **VB Scripting for CATIA V5**, is the most complete guide to teach you how to write macros for CATIA V5! Through a series of example codes and tutorials you'll learn how to unleash the full power and potential of CATIA V5. No programming experience is required! This text will cover the core items to help teach beginners important concepts needed to create custom CATIA macros. More importantly, you'll learn how to solve problems and what to do when you get stuck. Once you begin to see the patterns you'll be flying along on your own in no time.

Did You Like CATIA V5 Tips and Tricks?

Before you go, I'd like to say "thank you" for purchasing my little book. I know you could have picked up other guides or simply searched the internet, but you took a chance on my book. So a big thanks for ordering this book and reading all the way to the end.

Now I'd like to ask for a *small* favor. Could you please take a minute or two and leave a review for this book on Amazon? This feedback will help me continue to write the kind of books that help you get results. And if you loved it, then please let me know!

Appendix I: Symbol Shortcuts

The following is a list of shortcuts to generate symbols:

Alt + 0176 = ° (Degrees)
Alt + 0149 = • (Bullet)
Alt + 0162 = ¢
Alt + 0188 = ¼
Alt + 0189 = ½
Alt + 0190 = ¾
Alt + 0177 = ±
Alt + 0178 = ^{2}
Alt + 0179 = ^{3}
Alt + 0163 = £
Alt + 0128 = €
Alt + 0151 = — (m dash)
Alt + 0150 = – (n dash)
Alt + 0187 = »
Alt + 0169 = ©
Alt + 0174 = ®
Alt + 0165 = ¥
Alt + 0177 = ±
Alt + 0247 = ÷
Alt + 0166 = ¦
Alt + 0134 = †
Alt + 0227 = ã
Alt + 0191 = ¿
Alt + 0161 = ¡
Alt + 0209 = Ñ
Alt + 0241 = ñ
Alt + 0225 = á
Alt + 0233 = é
Alt + 0237 = í
Alt + 0243 = ó
Alt + 0250 = ú
Alt + 0252 = ü
Alt + 0186 = ° (1° = primero)
Alt + 0170 = ^{2} (2^2 = segunda)

Appendix II: Acronyms

The following terms are used throughout this text (in alphabetical order):

API: Application Programming Interface
BREP: Boundary Representation
CAA: CATIA Application Architecture
CAD: Computer Aided Design
CATDUA: CATIA Data Upward Assistant
CATIA: Computer Aided Three Dimensional Interactive Application
CLI: Command Line Interface
COM: Component Object Model
DLL: Dynamic Linked Library
DMU: Digital Mock-Up
DPI: Dots per inch
FBM: Feature Based Modeling
GUI: Graphical User Interface
GDT: Geometrical Dimensioning and Tolerancing
IDE: Integrated Development Environment
IDL: Interface Definition Language
MDB: Model Based Definition
OLE: Object Linking and Embedding
OOP: Object Oriented Programming
PBD: Product Based Definition
PDM: Product Data Management
PLM: Product Lifecycle Management
RADE: Rapid Application Development Environment
TLB: Type Library File
UUID: Universal Unique Identifier
VB6: Visual Basic 6
VBA: Visual Basic for Applications
VBE: Visual Basic Editor

Appendix III: Tools and Resources

Below is a list of tools and resources I personally use (or have used) to help me with CATIA, programming, putting together this eBook, and creating my website. I highly recommend each and every one (and wouldn't list it here if I didn't). Some of these premium services do come with a price (just being honest here), but seriously I wouldn't risk mentioning them here if I didn't know they work great and will save a lot of people time and money.

Where to Get Your CATIA Macro Questions Answered

COE (CATIA Operators Exchange): Post your questions in the Develop and Deploy forum to get answers from knowledgeable and professional CATIA users. http://www.coe.org/p/fo/si/topic=113

Eng Tips Forum: Post your CATScript questions in the CATIA Products forum and get great feedback from CATIA power users.

http://www.eng-tips.com/threadminder.cfm?pid=560

StackOverFlow: Use this site to ask questions when you get stuck and there are tons of knowledgeable programmers willing to help you out.

http://stackoverflow.com/

CATIA V5: The official Dassault Systèmes website: http://www.3ds.com/

3D XML Player: Download the free player to insert 3D models into your Office documents.

http://www.3ds.com/products/3dvia/3d-xml/1/

Contact Me: I'll do my best to respond to you in a timely matter but due to the number of emails I get daily I may or may not be able to. Please be patient.

http://www.scripting4v5.com/Contact-Us/

CATIA, Programming, and Other Books

For more information on CATIA, I recommend checking out these resources from Amazon.com:

CATIA V5 R20 for Designers: An extremely detailed and helpful book, not just for beginners but intermediate and advanced 3D modelers and engineers as well.

Introduction to CATIA V5 R19 (A hands-on Tutorial Approach): The tutorials seem to touch on every single icon that is available in CATIA and includes step-by-step tutorials with nice graphics.

27978038R00046

Made in the USA
San Bernardino, CA
18 December 2015